Studies in Space Policy

Volume 21

Series Editor
European Space Policy Institute, Vienna, Austria

Edited by: European Space Policy Institute, Vienna, Austria
Director: Jean-Jacques Tortora

Editorial Advisory Board:
Genevieve Fioraso
Gerd Gruppe
Pavel Kabat
Sergio Marchisio
Dominique Tilmans
Ene Ergma
Ingolf Schädler
Gilles Maquet
Jaime Silva

The use of outer space is of growing strategic and technological relevance. The development of robotic exploration to distant planets and bodies across the solar system, as well as pioneering human space exploration in earth orbit and of the moon, paved the way for ambitious long-term space exploration. Today, space exploration goes far beyond a merely technological endeavour, as its further development will have a tremendous social, cultural and economic impact. Space activities are entering an era in which contributions of the humanities—history, philosophy, anthropology—, the arts, and the social sciences—political science, economics, law—will become crucial for the future of space exploration. Space policy thus will gain in visibility and relevance. The series Studies in Space Policy shall become the European reference compilation edited by the leading institute in the field, the European Space Policy Institute. It will contain both monographs and collections dealing with their subjects in a transdisciplinary way. https://espi.or.at/about-us/governing-bodies

More information about this series at http://www.springer.com/series/8167

Annette Froehlich • Vincent Seffinga
Editors

The United Nations and Space Security

Conflicting Mandates Between UNCOPUOS and the CD

Editors
Annette Froehlich ⓘ
European Space Policy Institute
Vienna, Austria

Vincent Seffinga ⓘ
European Space Policy Institute
Vienna, Austria

ISSN 1868-5307 ISSN 1868-5315 (electronic)
Studies in Space Policy
ISBN 978-3-030-06024-4 ISBN 978-3-030-06025-1 (eBook)
https://doi.org/10.1007/978-3-030-06025-1

© Springer Nature Switzerland AG 2020
This work is subject to copyright. All rights are reserved by the Publisher, whether the whole or part of the material is concerned, specifically the rights of translation, reprinting, reuse of illustrations, recitation, broadcasting, reproduction on microfilms or in any other physical way, and transmission or information storage and retrieval, electronic adaptation, computer software, or by similar or dissimilar methodology now known or hereafter developed.
The use of general descriptive names, registered names, trademarks, service marks, etc. in this publication does not imply, even in the absence of a specific statement, that such names are exempt from the relevant protective laws and regulations and therefore free for general use.
The publisher, the authors, and the editors are safe to assume that the advice and information in this book are believed to be true and accurate at the date of publication. Neither the publisher nor the authors or the editors give a warranty, express or implied, with respect to the material contained herein or for any errors or omissions that may have been made. The publisher remains neutral with regard to jurisdictional claims in published maps and institutional affiliations.

This Springer imprint is published by the registered company Springer Nature Switzerland AG.
The registered company address is: Gewerbestrasse 11, 6330 Cham, Switzerland

Executive Summary

Outer space and issues pertaining to outer space are inherently international. The designation of outer space as the province of all mankind, the focus on cooperation between states and the recognition of the common interest of all mankind in the progress of the exploration and use of outer space for peaceful purposes included in various space law instruments illustrate the inherent international nature of space matters. The mandate given to the Committee on the Peaceful Uses of Outer Space (COPUOS) in United Nations General Assembly (UNGA) Resolution 1472 (XIV) stipulates that it should focus on the peaceful uses of outer space and the legal problems that arise from the exploration of outer space. Through UNGA Resolution S-10/2, the Conference on Disarmament (CD) was given the mandate to discuss the prevention of an arms race in outer space.

Even a cursory glance at the stipulated mandates reveals that the mandates do not necessarily encompass all uses of outer space. Rather, depending on the interpretation of the mandates, the military use of outer space can be discussed in either forum or in neither forum, leaving an apparent gap as to which forum is the locus for international discussion of the military uses of outer space. This question is the focal point of this publication. To answer this question, this book first determines the mandates given to COPUOS and the CD at their respective establishment. The initial mandates, however, leave ambiguity about the discussion of space matters internationally. Depending on the interpretation of the term 'peaceful', divergent interpretations of the mandates of the forums arise. This leads to a void in the discussion of space matters internationally, specifically about the 'non-arms military' use of outer space.

Thereafter, the development of the mandates is analysed from the establishment of the forums until the present. This analysis illustrates that the interpretation of the mandates has fluctuated. Divergent interpretations existed at the establishment of COPUOS in the late '50s. During the deliberations on the five UN Space Treaties, these interpretations seemed to change, with the interpretation that COPUOS was mandated to discuss the military use of outer space and the limitation of the use of outer space to ensure that it is used for exclusively peaceful purposes prevailing.

However, following the adoption of Resolution S-10/2, the diverging interpretations reappeared. In practice, both COPUOS and the CD exceed their mandate. COPUOS, on occasion, discusses military uses of outer space and disarmament matters. The CD frequently deliberates on matters that are 'non-arms military' uses of outer space and sometimes even peaceful uses of outer space (or at least have an impact on the peaceful use of outer space). The overlap between the discussions in both forums demonstrates the interrelated nature of space matters. The absence of any real cooperation then leads to the question whether the current UN space-related framework can effectively discuss these space matters.

The final chapter first briefly describes four space matters that are currently under discussion in COPUOS and the CD: the ways and means of maintaining outer space for peaceful purposes, space traffic management, space debris and the long-term sustainability of outer space activities. These issues are described as examples of space matters under consideration that cannot simply be divided into military and non-military or peaceful and aggressive use of outer space. Instead, the issues illustrate the interrelated nature of space matters. The interrelated nature of these space matters and the conclusion that they need to be dealt with in a cooperative effort between COPUOS and the CD is then used as the background for the evaluation of the current UN space-related framework. It is argued that the current UN space-related framework, in which space matters are discussed in the CD and COPUOS separately despite the interrelated nature of space matters, is not effective, first, because of the current deadlock in the CD that prevents the CD from making any substantive progress and, second, because a result or outcome on an issue in one forum affects the discussion in the other forum, which might lead to unforeseen complications or consequences for the further discussion of the issue. Two possible solutions are given to deal more effectively with space matters internationally, namely closer and consistent cooperation between the two forums or giving COPUOS a specific mandate to discuss the military use of outer space and disarmament matters within a specific context.

However, two important notes are placed. First, cooperation between the two forums can only effectively take place when the CD resolves its deadlock. Second, both solutions run the risk of falling in the same pitfall as the CD, namely that states cannot agree on how to proceed with the discussion of the disarmament aspects of certain space matters, which will then lead to a deadlock. This is a reasonable expectation because historic evidence indicates that the contentious and political nature of the topic of the military use of outer space leads to discussions on the topic coming to a standstill. The solutions are thus not without their problems. Nevertheless, one thing remains certain: space technology will continue to develop, and this development will bring with it new legal problems that will need to be addressed in a more timely and effective manner than is currently the case.

Contents

1	**Introduction**		1
	1.1	The Setting	1
	1.2	Approach of the Study	5
2	**Initial Mandates of the Committee on the Peaceful Uses of Outer Space (COPUOS) and the Conference on Disarmament (CD)**		7
	2.1	The Initial Mandate of the Committee on the Peaceful Uses of Outer Space (COPUOS)	9
	2.2	The Initial Mandate of the Conference on Disarmament (CD)	15
		2.2.1 Ten-Nation Committee	15
		2.2.2 Eighteen-Nation Committee	17
		2.2.3 Conference of the Committee on Disarmament	21
		2.2.4 The Conference on Disarmament	22
	2.3	The Discussion of Space Matters Internationally Based on the Initial Mandates of COPUOS and the CD	24
3	**The Development of the Mandates of the Committee on the Peaceful Uses of Outer Space (COPUOS) and the Conference on Disarmament (CD) and the Collaboration Between the Forums**		29
	3.1	Historical Background	31
		3.1.1 The USSR–U.S. Cold War and Space Race	32
		3.1.2 European Integration and Space Cooperation	35
		3.1.3 The People's Republic of China's Position in the UN	39
	3.2	The Development of the Mandate of Committee on the Peaceful Uses of Outer Space (COPUOS)	42
		3.2.1 The Development of the Mandate of COPUOS During the Negotiations of the UN Space Treaties	43
		3.2.2 The Development of the Mandate of COPUOS After the Five Space Treaties	78
	3.3	The Development of the Mandate of the Conference on Disarmament (CD)	94

		3.3.1	The Prevention of an Arms Race in Outer Space After Resolution S-10/2.......................................	95
		3.3.2	The Prevention of an Arms Race in Outer Space Since the Deadlock of the CD................................	99
	3.4	Interim Conclusion on the Mandates of COPUOS and the CD and the Collaboration Between COPUOS and the CD.........		102
4	**The Future of the UN Space-Related Framework**...............			107
	4.1	The Major Space-Related Issues Facing the International Community.......................................		108
		4.1.1	Ways and Means of Maintaining Outer Space for Peaceful Purposes..................................	108
		4.1.2	Space Traffic Management (STM).................	109
		4.1.3	Space Debris....................................	112
		4.1.4	Long-Term Sustainability of Outer Space Activities.....	114
	4.2	The Future Relevance of the UN Space-Related Framework....		115

About the Contributors

Annette Froehlich is a scientific expert seconded from the German Aerospace Center (DLR) to the European Space Policy Institute (Vienna), and Honorary Adjunct Senior Lecturer at the University of Cape Town (SA) at SpaceLab. She graduated in European and International Law at the University of Strasbourg (France), followed by business-oriented postgraduate studies and her PhD at the University of Vienna (Austria). Responsible for DLR and German representation to the United Nations and international organisations, Dr. Froehlich was also a member/alternate head of delegation of the German delegation of UNCOPUOS. Moreover, Dr. Annette Froehlich is the author of a multitude of specialist publications and serves as a lecturer at various universities worldwide in space policy, law and society aspects. Her main areas of scientific interest are European space policy, international and regional space law, emerging space countries, space security, and space and culture. She has also launched as editor the new scientific series "Southern Space Studies" (Springer publishing house) dedicated to Latin America and Africa.

Vincent Seffinga is a Ph.D. researcher at the Department of Law at the European University Institute (EUI) in Florence, Italy. He completed his undergraduate studies at the University Utrecht and his graduate studies at the University of Amsterdam. His research interests lie in the field of space law, with his work focusing specifically on the comprehensive regulation of outer space activities and space traffic management. In addition, he has written about national space legislation and has been awarded the Secure World Foundation Young Professionals Scholarship Award for attending the 2018 International Aeronautical Congress.

Ruiyan Qiu obtained her master's degree at the International Institute of Air and Space Law at Leiden University in 2018. Her study was fully sponsored by the CSC scholarship, which was granted by the Chinese government. She will start her traineeship at the International Civil Aviation Organization this year.

Abbreviations

ARRA	Agreement on the Rescue of Astronauts the Return of Astronauts and the Return of Objects Launched into Outer Space
ASAT	Anti-satellite weapons
CCD	Conference of the Committee on Disarmament
CD	Conference on Disarmament
CIA	Central Intelligence Agency
COPUOS	Committee on the Peaceful Uses of Outer Space
CTBTO	Comprehensive Nuclear-Test-Ban Treaty Organisation
DBS principles	Principles Governing the Use by States of Artificial Earth Satellites for International Direct Television Broadcasting
GDR	German Democratic Republic
EAEC	European Atomic Energy Community
EC	European Communities
ECSC	European Coal and Steel Community
EEC	European Economic Community
Eighteen-Nation Committee	Eighteen-Nation Committee on Disarmament
ELDO	European Launcher Development Organisation
EO	Earth Observation
ESA	European Space Agency
ESRO	European Space Research Organisation
EU	European Union
EUCoC	European Union Draft Code of Conduct for Outer Space Activities
First Committee	Disarmament and International Security Committee
Fourth Committee	Special Political and Decolonization Committee
GEERS	Groupe d'Études Européen pour la collaboration dans la domaine des recherches spatiales
IAA	International Academy of Astronautics

IADC	Inter-Agency Space Debris Coordination Committee
ISS	International Space Station
LIAB	Convention on International Liability for Damage Caused by Space Objects
Legal Subcommittee	Legal Subcommittee of the Committee on the Peaceful Uses of Outer Space
LTS	Long-term sustainability of outer space activities
MOON	Agreement Governing the Activities of States on the Moon and Other Celestial Bodies
NASA	National Aeronautics and Space Administration
NATO	North Atlantic Treaty Organisation
OST	Treaty on Principles Governing the Activities of States in the Exploration and Use of Outer Space, including the Moon and Other Celestial Bodies
PAROS	Prevention of an Arms Race in Outer Space Treaty
PPWT	Treaty on the Prevention of the Placement of Weapons in Outer Space
PRC	People's Republic of China
PTBT	Treaty Banning Nuclear Tests in the Atmosphere, in Outer Space and Under Water
REG	Convention on Registration of Objects Launched into Outer Space
ROC	Republic of China (Taiwan)
RS principles	Principles Relating to Remote Sensing of the Earth from Space
SPC	Special Political Committee of the United Nations
SSA	Space Situational Awareness
STM	Space Traffic Management
STSC	Scientific and Technical Subcommittee of the Committee on the Peaceful Uses of Outer Space
Ten-Nation Committee	Ten-Nation Committee on Disarmament
UAR	United Arab Republic
UK	United Kingdom of Great Britain and Northern Ireland
UN	United Nations
UN Charter	Charter of the United Nations
UNGA	United Nations General Assembly
UNSC	United Nations Security Council
UN-Space	Inter-Agency Meeting on Outer Space Activities
U.S.	United States of America
USSR	Union of Soviet Socialist Republics

Chapter 1
Introduction

Annette Froehlich, Vincent Seffinga, and Ruiyan Qiu

Abstract An examination of the mandates of the Committee on the Peaceful Uses of Outer Space (COPUOS) and the Conference on Disarmament (CD) reveals that their individual mandates do not necessarily encompass all uses of outer space. Rather, depending on the interpretation of their mandates, the military use of outer space can be discussed in either forum or in neither. This begs the question in which forum the military uses of outer space are discussed internationally. This question is the focal point of this book. This introduction outlines the importance of this question by describing the extent of the military use of outer space. Furthermore, it lays out the structure of the research, which is divided into three parts. First, the research will examine the initial mandates given to COPUOS and the CD. Second, the research will analyse the further development and practical interpretation of those mandates. Finally, the research will assess the effectiveness of the current UN space-related framework to cope with near to medium future space matters that affect both the military and non-military use of outer space.

1.1 The Setting

With the creation of the Committee on the Peaceful Uses of Outer Space (COPUOS) in 1959, the international community established a forum for the discussion of issues related to outer space activities. The name of the committee implies that the mandate of COPUOS is limited to the discussion of issues pertaining to the peaceful uses of

---

A. Froehlich (✉) · V. Seffinga
European Space Policy Institute, Vienna, Austria
e-mail: annette.froehlich@espi.or.at

R. Qiu
International Institute of Air and Space Law, Leiden University, Leiden, Netherlands

outer space.¹ This presumption is corroborated by the mandate stipulated in United Nations General Assembly (UNGA) Resolution 1472 (XIV), in which COPUOS was established

> (a) To review, as appropriate, the area of international co-operation, and to study practical and feasible means for giving effect to programmes in the ***peaceful uses (emphasis added)*** of outer space which could appropriately be undertaken under United Nations auspices, including, *inter alia*:
>
> > (i) Assistance for the continuation on a permanent basis of the research on outer space carried on within the framework of the International Geophysical Year;
> > (ii) Organization of the mutual exchange and dissemination of information on outer space research;
> > (iii) Encouragement of national research programmes for the study of outer space, and the rendering of all possible assistance and help towards their realization;
>
> (b) To study the nature of legal problems which may arise from the ***exploration (emphasis added)*** of outer space;[...].²

This mandate is a reiteration of the mandate given to the *ad hoc* Committee on the Peaceful Uses of Outer Space in UNGA Resolution 1348 (XIII).³ Thus, a cursory glance indicates that COPUOS focuses exclusively on the *peaceful uses* of outer space and legal problems with respect to the *exploration* of outer space. Within the United Nations (UN) framework, COPUOS reports to the Special Political and Decolonization Committee of the UN General Assembly (Fourth Committee).

However, the use of outer space for military purposes has been, and still is, an important part of the activities conducted in outer space.⁴ The military uses of outer space that are lawful are limited through Article III OST,⁵ Article IV OST and Article 3 MOON.⁶ Article III OST stipulates that activities in outer space need to be carried on 'in accordance with international law, including the Charter of the United Nations'. Therefore, both Article 2(4) UN Charter, stipulating on the prohibition on the threat or use of force, and Article 51 UN Charter, stipulating on the right to self-defence, are applicable to outer space.⁷ Although outer space or space-based assets

---

¹Peter Jankowitsch, 'The Background and History of Space Law' in Frans von der Dunk & Fabio Tronchetti (eds), *Handbook of Space Law* (Edward Elgar Publishing 2015) 18.

²UNGA Res 1472 (XIV) (12 December 1959).

³UNGA Res 1348 (XIII) (13 December 1958).

⁴Fabio Tronchetti, 'Legal Aspects of the Military Uses of Outer Space' in Frans von der Dunk & Fabio Tronchetti (eds), *Handbook of Space Law* (Edward Elgar Publishing 2015) 331 | David A. Koplow, 'The Fault Is Not in Our Stars: Avoiding an Arms Race in Outer Space' (2018) 59 Harvard International Law Journal 331, 335–336 | Anél Ferreira-Snyman, 'Selected Legal Challenges Relating to the Military Use of Outer Space, With Specific Reference to Article VI of the Outer Space Treaty' (2015) 18 Potchefstroom Electronic Law Journal 488, 495–496.

⁵Treaty on Principles Governing the Activities of States in the Exploration and Use of Outer Space, including the Moon and Other Celestial Bodies (adopted 19 December 1966, entered into force 10 October 1967) 610 UNTS 205 [OST].

⁶Agreement Governing the Activities of States on the Moon and Other Celestial Bodies (adopted 18 December 1979, entered into force 11 July 1984) 1363 UNTS 3 [MOON].

⁷Charter of the United Nations (adopted 26 June 1945, entered into force 24 October 1945) [UN Charter] | Steven Freeland, 'In Heaven as on Earth? The International Legal Regulation of

thus cannot be used in a manner that constitutes a 'threat or use of force', Article 51 UN Charter acknowledges that states can make use of outer space and space-based assets in their self-defence.[8] Of course, the applicability of these articles brings with it the extensive discussion under general international law on the meaning, scope, application of, and jurisprudence on, these articles.

Article IV OST and Article 3 MOON contain limitations more specific to outer space. In essence, Article IV OST prohibits any objects carrying nuclear weapons or other kinds of weapons of mass destruction to be placed in orbit, installed on celestial bodies or otherwise stationed in outer space.[9] Furthermore, Article IV OST forbids establishing military bases, installations and fortifications; testing any type of weapons; and conducting military manoeuvres on celestial bodies.[10] Article 3 MOON expands on this by stipulating that 'the Moon shall be used by all States Parties exclusively for peaceful purposes', which in accordance with Article 1 MOON extends to all celestial bodies within the solar system. Moreover, the MOON prohibits any threat or use of force, or any hostile act or threat of hostile act on the Moon. Nevertheless, these provisions merely stipulate a limited prohibition on the military use of outer space; military satellites or conventional weapons, for example, can still be placed into orbit.[11]

The fact that certain military uses of outer space are lawful but that the mandate of COPUOS is explicitly limited to peaceful uses of outer space leads to uncertainty with respect to where some of the military uses of outer space should be discussed

---

the Military Use of Outer Space' (2011) 8 U.S.-China Law Review 272, 276–277 | Francis Grimal & Jae Sundaram, 'The Incremental Militarization of Outer Space: A Threshold Analysis' (2018) 17 Chinese Journal of International Law 45, 55 | Anél Ferreira-Snyman, 'Selected Legal Challenges Relating to the Military Use of Outer Space, With Specific Reference to Article VI of the Outer Space Treaty' (2015) 18 Potchefstroom Electronic Law Journal 488, 494–495.

[8]Steven Freeland, 'In Heaven as on Earth? The International Legal Regulation of the Military Use of Outer Space' (2011) 8 U.S.-China Law Review 272, 276–277 | Francis Grimal & Jae Sundaram, 'The Incremental Militarization of Outer Space: A Threshold Analysis' (2018) 17 Chinese Journal of International Law 45, 55 | Anél Ferreira-Snyman, 'Selected Legal Challenges Relating to the Military Use of Outer Space, With Specific Reference to Article VI of the Outer Space Treaty' (2015) 18 Potchefstroom Electronic Law Journal 488, 494–495.

[9]David A. Koplow, 'The Fault Is Not in Our Stars: Avoiding an Arms Race in Outer Space' (2018) 59 Harvard International Law Journal 331, 348 | Francis Grimal & Jae Sundaram, 'The Incremental Militarization of Outer Space: A Threshold Analysis' (2018) 17 Chinese Journal of International Law 45, 55.

[10]David A. Koplow, 'The Fault Is Not in Our Stars: Avoiding an Arms Race in Outer Space' (2018) 59 Harvard International Law Journal 331, 348 | Francis Grimal & Jae Sundaram, 'The Incremental Militarization of Outer Space: A Threshold Analysis' (2018) 17 Chinese Journal of International Law 45, 55.

[11]Fabio Tronchetti, 'Legal Aspects of the Military Uses of Outer Space' in Frans von der Dunk & Fabio Tronchetti (eds), *Handbook of Space Law* (Edward Elgar Publishing 2015) 338 | Anél Ferreira-Snyman, 'Selected Legal Challenges Relating to the Military Use of Outer Space, With Specific Reference to Article VI of the Outer Space Treaty' (2015) 18 Potchefstroom Electronic Law Journal 488, 495–496 | Francis Grimal & Jae Sundaram, 'The Incremental Militarization of Outer Space: A Threshold Analysis' (2018) 17 Chinese Journal of International Law 45, 48.

internationally. Historically, it has become clear that the discussion on the disarmament of outer space takes place in the UN disarmament system. The Partial Test-Ban Treaty (PTBT), for example, was (partially) negotiated in the Eighteen-Nation Committee on Disarmament (Eighteen-Nation Committee).[12] Subsequently, the Committee on Disarmament, as the predecessor of the Conference on Disarmament (CD), which reports to the Disarmament and International Security Committee of the UN General Assembly (First Committee), put the prevention of an arms race in outer space on its agenda 'in accordance with the spirit of the Outer Space Treaty' per the recommendation of the Final Document of the Tenth Special Session of the General Assembly.[13] The CD has since then addressed the prevention of an arms race in outer space through initiatives such as the Treaty on the Prevention of the Placement of Weapons in Outer Space (PPWT) and a general Prevention of an Arms Race in Outer Space (PAROS) Treaty. Although neither the PPWT nor a general PAROS Treaty has been realised, the CD nevertheless aims to realise a legally binding international instrument on the prevention of an arms race in outer space.[14]

However, the inclusion of this agenda item on the CD agenda does not give complete certainty as to where all military uses of outer space are to be discussed internationally. The disarmament of outer space and the prevention of an arms race in outer space are only *aspects* of the military use of outer space. There is a wealth of non-arms military uses of outer space: telecommunications, remote sensing, GPS, not to mention the inherent dual-use nature of much space technology.[15]

Moreover, there are efforts to increase the military use of outer space. The prime example of such efforts is the U.S. plan for a U.S. Space Force.[16] Furthermore, although both COPUOS and the CD address issues pertaining to the use of outer space, coordination and collaboration between the two forums has been minimal. The matters discussed in the two forums are becoming increasingly more related. For example, the long-term sustainability of outer space activities (LTS), space debris

---

[12]Treaty Banning Nuclear Tests in the Atmosphere, in Outer Space and Under Water (adopted 5 August 1963, entered into force 10 October 1963) 480 UNTS 43 [PTBT].

[13]UNGA Res S-10/2 (28 June 1978) UN Doc A/RES/S-10/2, par. 80 | UNGA 'Thirty-Sixth Session Report of the Disarmament Commission' UN GAOR 36th Session Supp No 42 UN Doc A/36/42 (1981), par. 19.

[14]UNGA First Committee (72nd Session) 'Draft Resolution: Further Practical Measures for the Prevention of an Arms Race in Outer Space' (13 October 2017) UN Doc A/C.1/72/L.54.

[15]David A. Koplow, 'The Fault Is Not in Our Stars: Avoiding an Arms Race in Outer Space' (2018) 59 Harvard International Law Journal 331, 335–336 | Anél Ferreira-Snyman, 'Selected Legal Challenges Relating to the Military Use of Outer Space, With Specific Reference to Article VI of the Outer Space Treaty' (2015) 18 Potchefstroom Electronic Law Journal 488, 495–496.

[16]Steven Freeland, 'The US Plan for a Space Force Risks Escalating a "Space Arms Race"' (*The Conversation*, 10 August 2018) <http://theconversation.com/the-us-plan-for-a-space-force-risks-escalating-a-space-arms-race-101368> accessed 6 December 2018 | Melissa de Zwart, 'The International Context of Trump's Space Force' (*Australian Institute of International Affairs*, 23 June 2018) <http://www.internationalaffairs.org.au/australianoutlook/international-context-trump-space-force/> accessed 6 December 2018 | Babak Shakouri Hassanabadi, 'Space Force and International Space Law' (*The Space Review*, 30 July 2018) <http://www.thespacereview.com/article/3543/1> accessed 6 December 2018.

mitigation and removal, and space traffic management (STM) will all need to be addressed by both forums. This has been acknowledged by the UNGA through the convening of a joint panel discussion of the First and Fourth Committees on possible challenges to space security and sustainability.[17] Nevertheless, further collaboration or interaction has not been realised.

In light of the aforementioned, this research will examine the development of the mandate of both COPUOS and the CD from a historical perspective. Furthermore, the research will anticipate likely major space-related matters that will have to be debated in both forums in the near to medium-term future and evaluate the effectiveness of the current system to cope with these issues. The rationale behind the research is thus to analyse the current system of discussing space matters internationally and to examine if and how this system can be improved. This report will detail the history and development of the mandates of COPUOS and the CD and the coordination and cooperation between the forums. Furthermore, it will have a future-oriented perspective on the likely challenges that these forums, out of necessity, will have to deal with, and conclude with observations on which way the current system can change to be able to cope with the evolution of space matters. This is necessary for the proper discussion of space matters because such space developments cannot simply be divided between peaceful and disarmament matters—they are interrelated.

1.2 Approach of the Study

The research will be divided into three chapters, which will each have a specific phase of the development of the current system as their focal point. The first chapter will concentrate on the creation of the current system of the UN that features two separate forums that each have a mandate to discuss different specific space-related matters. Therefore, this first chapter will examine the establishment and the initial mandates of COPUOS and the CD (as it relates to outer space) and how those initial mandates could be interpreted. With respect to COPUOS, the chapter will examine the period from the inception of the *ad hoc* Committee on the Peaceful Uses of Outer Space in 1958 until COPUOS was fully established through UNGA Resolution 1472 (XIV). Likewise, the chapter will examine the mandate of the CD to discuss the disarmament of outer space and the prevention of an arms race in outer space. It will follow the development of this subject in the disarmament framework, including the CD's predecessors. Thus, the first chapter will discuss the responsibilities that have been bestowed on COPUOS and the CD, *i.e.* which issues are discussed in which forum, and the purpose and goal of the forums.

The second chapter will analyse the subsequent evolution and development of the mandates of the forums from the moment that COPUOS was established and the CD received the mandate to discuss the prevention of an arms race in outer space. The

---

[17]UNGA Res 71/90 (6 December 2016) UN Doc A/RES/71/90, par. 15.

period from the early 1960s (COPUOS) and the early 1980s (CD) until the present will be examined. The analysis will be situated in the historical factors that contributed to the development of the forums and will include an assessment of the extent to which the space-related UN framework has evolved—or has not evolved—along with the changing global context of space issues. This analysis will assess the extent to which the current arrangements for the discussion of space matters internationally are different from the vision and expectations that were put forward during the creation of COPUOS and mandating of the CD. Primary sources will be used as much as feasible. For example, the statements made by states with respect to the relevant resolutions or legal instruments, such as the resolutions specifying the mandate of the committees and discussions on the articles in the UN Space Treaties that deal with military/non-military issues, will be examined. Secondary sources, such as scholarly contributions, will be used to complement the statements made by states, when necessary.

The third chapter will have a forward-looking perspective. First, it will outline the major space-related matters that are currently being debated in either of the forums but are a concern for both forums, namely LTS, space debris and space traffic management (STM). These issues will be briefly explained to create the required context. Second, the adequacy of the current system to cope with these issues will be evaluated. For example, can an issue such as the long-term sustainability of outer space activities be adequately addressed when it is discussed in two separate forums, instead of one centralised forum, or through a cooperative effort? It will then assess the extent to which the rationale that prevailed at the time of the establishment of the current framework for debating space matters internationally is still relevant. Moreover, it will examine whether changes to the status quo are necessary to ensure the proper discussion of space issues in the light of the intertwined relationship between military and non-military uses of outer space. The analysis in this chapter, both with respect to anticipating the major space-related issues and the adequacy of the current system, will be conducted on the basis of the findings in the previous chapter and scholarly publications.

Chapter 2
Initial Mandates of the Committee on the Peaceful Uses of Outer Space (COPUOS) and the Conference on Disarmament (CD)

Annette Froehlich, Vincent Seffinga, and Ruiyan Qiu

Abstract On a state level, space matters are discussed globally in two separate forums: the Committee on the Peaceful Uses of Outer Space (COPUOS), which deals with the peaceful uses of outer space, and the Conference on Disarmament (CD), which focuses on the prevention of an arms race in outer space and other matters pertaining to the disarmament of outer space. COPUOS was established in 1959 through Resolution 1348 (XIII), while the CD was given the official mandate to discuss the prevention of an arms race in outer space in 1981 through Resolution S-10/2. This chapter examines the mandates established in these resolutions and the various interpretations of those mandates over time. First, this chapter examines and evaluates the mandate given to COPUOS, coming to the conclusion that the mandate is ambiguous because the term 'peaceful uses' is not defined. Thereafter, this chapter discusses the evolution of the disarmament framework, including the CD's predecessors, and the eventual mandate given to the CD, which appears better defined and less ambiguous. Finally, the chapter considers the mandates given to the two forums and determines the effect that this has had on the discussion of space matters internationally within the United Nations (UN) space-related framework. It concludes that there appears to be a void in the discussion of space matters internationally, specifically with regard to the 'non-arms military' use of outer space, because of divergent interpretations of the mandates of the forums by key spacefaring states.

This chapter will examine the decision to discuss, or rather the eventuality of having to discuss, space matters internationally in two separate forums: the Committee on the Peaceful Uses of Outer Space (COPUOS), which deals with the peaceful uses of

A. Froehlich (✉) · V. Seffinga
European Space Policy Institute, Vienna, Austria
e-mail: annette.froehlich@espi.or.at

R. Qiu
International Institute of Air and Space Law, Leiden University, Leiden, Netherlands

© Springer Nature Switzerland AG 2020
A. Froehlich, V. Seffinga (eds.), *The United Nations and Space Security*,
Studies in Space Policy 21, https://doi.org/10.1007/978-3-030-06025-1_2

outer space, and the Conference on Disarmament (CD), which deals with the prevention of an arms race in outer space and other disarmament of outer space matters. Although COPUOS was established in 1958 as an *ad hoc* Committee through Resolution 1348 (XIII),[1] the CD was only given the formal mandate to discuss the prevention of an arms race in outer space in 1981.[2] The research will assess developments during this period to accurately study the inception of the current United Nations (UN) framework for the international discussion of space matters. First, this chapter will describe and evaluate the initial mandate given to COPUOS (Sect. 2.1). Thereafter, the evolution of the disarmament framework, including the CD's predecessors, and the eventual mandate given to the CD (Sect. 2.2), will be addressed. Finally, the chapter will consider the mandates given to the forums and determine the effect that this has on the discussion of space matters internationally through an analysis of the interpretation of the mandates (Sect. 2.3).

International discussion of space matters, at least within the UN framework, started in earnest through United Nations General Assembly (UNGA) Resolution 1148 (XII).[3] In paragraph 1(f), the resolution urges as follows:

> (…) that the States concerned, and particularly those which are members of the Sub-Committee of the Disarmament Commission, give priority to reaching a disarmament agreement which, upon its entry into force, will provide for the following:
>
> (…)
>
> (f) The joint study of an inspection system designed to ensure that the sending of objects through outer space shall be exclusively for peaceful and scientific purposes; (…).

This reflects the draft resolution recommended by the First Committee in its report, which was largely based on a draft resolution submitted by a large group of co-sponsors, consisting of Argentina, Australia, Belgium, Brazil, Canada, Chile, Colombia, Cuba, the Dominican Republic, Ecuador, France, Honduras, Italy, Laos, Liberia, the Netherlands, Nicaragua, Panama, Paraguay, Peru, the Philippines, Tunisia, the United Kingdom (UK) and the United States of America (U.S.).[4] The international discussion of space matters thus began as a part of the disarmament agenda and was located in the disarmament commission. Considering that the U.S. and the Union of Soviet Socialist Republics (USSR) held conflicting views on how to approach disarmament, it is no surprise that most of the discussion with respect to the resolution focused on other disarmament issues. Nevertheless, certain statements referred to non-disarmament space matters. The Philippines, for example,

---

[1] UNGA Res 1348 (XIII) (13 December 1958).
[2] UNGA Res S-10/2 (28 June 1978) UN Doc A/RES/S-10/2, par. 80 I UNGA 'Thirty-Sixth Session Report of the Disarmament Commission' UN GAOR 42nd Session Supp No 42 UN Doc A/36/42 (1981), par. 19.
[3] UNGA Resolution 1148 (XII) (14 November 1957).
[4] UNGA First Committee (12th Session) 'Regulation, Limitation and Balanced Reduction of All Armed Forces and All Armaments; Conclusion of an International Convention (Treaty) on the Reduction of Armaments and the Prohibition of Atomic, Hydrogen and Other Weapons of Mass Destruction' (11 November 1957) UN Doc A/3792, 4–5.

noted that space activities could cause international incidents and international disasters.[5] The Philippines also requested the Disarmament Commission, or the UNGA, to explore how to protect the Earth in light of the scientific advances with respect to outer space.[6] Accordingly, international attention to these space matters began in the Disarmament Commission.

It is important to note that Resolution 1148 (XII) already utilised the phrase 'exclusively for peaceful and scientific purposes', thereby seemingly indicating a limitation on the uses of outer space. Nevertheless, the initiative to have a separate forum to discuss space matters internationally occurred promptly after the adoption of Resolution 1148 (XII). Both draft resolutions submitted by the USSR and by the Twenty-Powers stipulated the establishment of a committee on the peaceful uses of outer space.[7]

2.1 The Initial Mandate of the Committee on the Peaceful Uses of Outer Space (COPUOS)

The push towards a specialised committee to address space matters internationally started through the request of the USSR to include on the agenda of the 13th session of the UNGA the question of 'The banning of the use of cosmic space for military purposes, the elimination of foreign military bases on the territories of other countries and international co-operation in the study of cosmic space'.[8] During the same session, the U.S. requested the inclusion on the agenda of an item called 'Programme for International Co-operation in the Field of Outer Space'.[9]

---

[5]UNGA 'Regulation, Limitation and Balanced Reduction of All Armed Forces and All Armaments; Conclusion of an International Convention (Treaty) on the Reduction of Armaments and the Prohibition of Atomic, Hydrogen and Other Weapons of Mass Destruction' UN GAOR 12th Session UN Doc A/PV.715 (14 November 1957), 449.

[6]UNGA 'Regulation, Limitation and Balanced Reduction of All Armed Forces and All Armaments; Conclusion of an International Convention (Treaty) on the Reduction of Armaments and the Prohibition of Atomic, Hydrogen and Other Weapons of Mass Destruction' UN GAOR 12th Session UN Doc A/PV.715 (14 November 1957), 449.

[7]UNGA 'Union of Soviet Socialist Republics: Draft Resolution' UN GAOR 13th Session UN Doc A/C.1/L.219 and Rev.1 (7 November 1958) | UNGA 'Australia, Belgium, Bolivia, Canada, Denmark, France, Guatemala, Ireland, Italy, Japan, Nepal, Netherlands, New Zealand, Sweden, Turkey, Union of South Africa, United Kingdom of Great Britain and Northern Ireland, United States of America, Uruguay and Venezuela' UN GAOR 13th Session UN Doc A/C.1/L.220 (13 November 1958).

[8]UNGA 'Union of Soviet Socialist Republics: Request for the Inclusion of An Item in the Provisional Agenda of the Thirteenth Session' UN GAOR 13th Session UN Doc A/3818 (17 March 1958).

[9]UNGA 'United States of America: Request for the Inclusion of an Additional Item in the Agenda of the Thirteenth Session' UN GAOR 13th Session UN Doc A/3902 (2 September 1958).

The difference between the phrasing of the proposed agenda items is apparent, with the USSR proposal focusing on the prohibition of military uses of outer space, while the U.S. proposal only addressed international cooperation. This difference becomes even more apparent in the explanatory memoranda attached to the requests. The USSR referred to a statement made by U.S. President Dwight Eisenhower pertaining to the banning of the use of outer space for military purposes and subsequently argued that this ban was merely focused on the prohibiting intercontinental ballistic rockets.[10] In addition, the USSR stated that the payload that a rocket carries was the factor that determined whether it was used for peaceful or military purposes. Therefore, the USSR argued that there should not be a ban on intercontinental ballistic rockets but rather that atomic and hydrogen bombs should be banned.[11] An important aspect of the USSR request is that it not only focused on the use of outer space, military or peaceful, but it connected the issue of the military use of outer space to the elimination of foreign military bases on the territories of other states.[12] The inclusion of this point related to the threat the USSR perceived from U.S. bases in Europe and Northern Africa, going so far as to state as follows:

> One cannot fail to see that, in raising the question of banning the use of cosmic space for military purposes, the United States is making an attempt, through a ban of the intercontinental ballistic rocket, to ward off a retaliatory nuclear blow through cosmic space while maintaining its numerous military bases on foreign territories intended for attacking with nuclear weapons the Soviet Union and the peaceful States friendly to it.[13]

As the USSR had proposed a complete ban of military uses of outer space, the request automatically concerned the use of outer space for peaceful purposes. The U.S. took a different approach, stating in its explanatory memorandum that outer space could be used for both destructive and peaceful purposes and that urgent steps were required to develop the peaceful uses of outer space, while parallel steps could be taken to reach agreements on the disarmament aspects of outer space.[14] Therefore, the U.S. saw discussion on the disarmament of outer space and discussion on the peaceful uses of outer space as distinct, stating as follows:

---

[10] UNGA 'Union of Soviet Socialist Republics: Request for the Inclusion of An Item in the Provisional Agenda of the Thirteenth Session' UN GAOR 13th Session UN Doc A/3818 (17 March 1958), par. 6.

[11] UNGA 'Union of Soviet Socialist Republics: Request for the Inclusion of An Item in the Provisional Agenda of the Thirteenth Session' UN GAOR 13th Session UN Doc A/3818 (17 March 1958), par. 7.

[12] UNGA 'Union of Soviet Socialist Republics: Request for the Inclusion of An Item in the Provisional Agenda of the Thirteenth Session' UN GAOR 13th Session UN Doc A/3818 (17 March 1958), par. 15.

[13] UNGA 'Union of Soviet Socialist Republics: Request for the Inclusion of An Item in the Provisional Agenda of the Thirteenth Session' UN GAOR 13th Session UN Doc A/3818 (17 March 1958), par. 12.

[14] UNGA 'United States of America: Request for the Inclusion of an Additional Item in the Agenda of the Thirteenth Session' UN GAOR 13th Session UN Doc A/3902 (2 September 1958), par. 3.

The General Assembly, (...), should begin to make the necessary steps to further those interests by declaring itself on the separability of the question of the peaceful uses of outer space from that of disarmament, (...) and by preparing for further careful consideration of this vital but complex matter through the establishment of a representative ad hoc committee (...).[15]

Accordingly, the different approaches towards the handling of space matters internationally were already apparent in the initial stages of the establishment of COPUOS. The difference of opinion between the two states led to two separate draft resolutions on how to proceed. The initial USSR draft resolution followed the rhetoric employed in its request to add the item to the agenda, namely by including the elimination of foreign military bases.[16] However, in the revised draft, the USSR removed this point and focused on the establishment of a committee for cooperation for the study of outer space for peaceful purposes, recommended a preparatory group to draft the rules and programme of the committee and proposed certain functions of the committee.[17] This revised draft resolution was the result of the discussions in the First Committee, where the difference in approach was recognised. As stated by the USSR:

Realizing that at the present time the United States and the other Western Powers refuse altogether to discuss the question of banning the use of cosmic space for military purposes, and seeking to meet the wishes of the many countries which are interested in the development of international co-operation in matters concerning the peaceful conquest of cosmic space, the USSR delegation took an important step in the direction of narrowing the gulf between the different positions and achieving agreement on at least one question, namely, international co-operation in the peaceful use of outer space.[18]

This brought the revised draft resolution closer in line with the draft resolution submitted by the group of 20 states, which did not attempt to ban the military use of outer space.[19] Instead, it emphasised the common aim that outer space should be used for peaceful purposes only.

The draft resolutions submitted by the USSR and the Twenty-Powers also differed in the mandate to be provided to the committee. The USSR resolution did not establish a committee but tried to establish a working group to establish such a committee. Therefore, the USSR draft resolution did not stipulate a mandate but recommended that the committee should have the functions of continuing the space

---

[15] UNGA 'United States of America: Request for the Inclusion of an Additional Item in the Agenda of the Thirteenth Session' UN GAOR 13th Session UN Doc A/3902 (2 September 1958), par. 4.

[16] UNGA 'Union of Soviet Socialist Republics: Draft Resolution' UN GAOR 13th Session UN Doc A/C.1/L.219 (7 November 1958).

[17] UNGA 'Union of Soviet Socialist Republics: Revised Draft Resolution' UN GAOR 13th Session UN Doc A/C.1/L.219/REV.1 (18 November 1958).

[18] UNGA 'Report of the First Committee' UN GAOR 13th Session UN Doc A/4009 (13 December 1958), par. 120.

[19] UNGA 'Australia, Belgium, Bolivia, Canada, Denmark, France, Guatemala, Ireland, Italy, Japan, Nepal, Netherlands, New Zealand, Sweden, Turkey, Union of South Africa, United Kingdom of Great Britain and Northern Ireland, United States of America, Uruguay and Venezuela' UN GAOR 13th Session UN Doc A/C.1/L.220 (13 November 1958).

research carried on within the framework of the International Geophysical Year, the mutual exchange and dissemination of information on space research and the coordination of national space research programmes.[20] In contrast, the Twenty-Powers Resolution sought to immediately establish an *ad hoc* committee that had to report on the activities and resources of the UN relating to the peaceful uses of outer space, study the area of international cooperation and the programmes in the peaceful uses of outer space that could be undertaken, study the future organisational arrangements to facilitate international cooperation and study the nature of legal problems that may arise from the exploration of outer space.[21]

These competing draft resolutions were discussed in the First Committee, with the result that the USSR resolution was not put to a vote, and the Twenty-Powers Resolution was adopted by a large majority.[22] The reason that such a large majority of states favoured the Twenty-Powers Resolution partly lay in the fact that, as was stated in the report, 'Other representatives urged that these military aspects should be considered within the framework of disarmament'.[23]

The subsequent discussion in the UNGA of the draft resolution adopted by the First Committee once more saw the discussion focus on the different positions taken by the U.S. and the USSR. The USSR referred to the discussion in the First Committee and noted that 'an overwhelming majority of the countries (...) quite clearly expressed their interest in ensuring the exclusively peaceful use of outer space'.[24] However, the USSR further noted that 'the United States flatly refused to consider the military aspects of the problem of outer space' and that 'the United States and the other Western Powers refuse altogether to discuss the question of banning the use of cosmic space for military purposes'.[25] The American delegate responded to those points by stating that when the USSR 'wishes to talk about realistic measures to ban the use of outer space for military purposes, the United States is ready' but that such an attempt had not been made.[26] Despite the different

---

[20] UNGA 'Union of Soviet Socialist Republics: Revised Draft Resolution' UN GAOR 13th Session UN Doc A/C.1/L.219/REV.1 (18 November 1958), par. 3.

[21] UNGA 'Australia, Belgium, Bolivia, Canada, Denmark, France, Guatemala, Ireland, Italy, Japan, Nepal, Netherlands, New Zealand, Sweden, Turkey, Union of South Africa, United Kingdom of Great Britain and Northern Ireland, United States of America, Uruguay and Venezuela' UN GAOR 13th Session UN Doc A/C.1/L.220 (13 November 1958), par. 1.

[22] UNGA 'Report of the First Committee' UN GAOR 13th Session UN Doc A/4009 (13 December 1958), par. 112–113.

[23] UNGA 'Report of the First Committee' UN GAOR 13th Session UN Doc A/4009 (13 December 1958), par. 107.

[24] UNGA 'Report of the First Committee' UN GAOR 13th Session UN Doc A/4009 (13 December 1958), par. 115.

[25] UNGA 'Report of the First Committee' UN GAOR 13th Session UN Doc A/4009 (13 December 1958), par. 117 and 120.

[26] UNGA 'Report of the First Committee' UN GAOR 13th Session UN Doc A/4009 (13 December 1958), par. 136.

positions, the Twenty-Powers Resolution was adopted by the UNGA as Resolution 1348 (XIII).

In accordance with the discussions conducted in the First Committee and the UNGA, Resolution 1348 (XIII) does not use the term 'military' but uses the term 'peaceful',[27] thereby keeping the military use of outer space lawful. The content of the resolution stipulates the mandate given to the *ad hoc* Committee on the Peaceful Uses of Outer Space. In subparagraphs 1(a), 1(b) and 1(c) the focus is on international cooperation in the peaceful uses of outer space. The *ad hoc* Committee is requested to assess the then-current resources of the UN and international bodies relating to the peaceful uses of outer space, the area of international cooperation and programmes in the peaceful uses of outer space that could appropriately be undertaken under UN auspices and the future organisational arrangements to facilitate international cooperation. Furthermore, the *ad hoc* Committee is requested to report on 'The nature of legal problems which may arise in the carrying out of programmes to explore outer space' in subparagraph 1(d). In light of the aforementioned mandate, and in light of the deliberations before the adoption of the resolution, the *ad hoc* Committee was limited to the discussion of peaceful uses of outer space. The term *peaceful* within this resolution was understood, at least by the states that sponsored the initial draft resolution, to exclude disarmament issues.

As a result, the *ad hoc* Committee began deliberating on space matters under the mandate stipulated under Resolution 1348 (XIII). In this discussion, it was reiterated that Resolution 1348 (XIII) exclusively referred to the peaceful uses of outer space.[28] The UK, for example, stated that '[COPUOS] was concerned only with the peaceful uses of outer space; and that excluded questions relating to other than peaceful uses and to disarmament'.[29] This was supported by Belgium, which stated that '(...) the matter being considered exclusively from the point of view of the peaceful uses of outer space'.[30] Accordingly, this view was reiterated in the report of the *ad hoc* Committee.[31]

The discussion on establishing COPUOS continued in the First Committee, wherein multiple states made statements with respect to the mandate of COPUOS. First, the U.S. stated that matters relating to the peaceful uses of outer space and the general problem of disarmament should be treated separately.[32] Second, an

---

[27]UNGA Res 1348 (XIII) (13 December 1958).

[28]*Ad hoc* COPUOS 'Report of the Working Group to the Legal Committee' UN GAOR 14th Session UN Doc A/AC.98/C.2/L.1 (9 June 1959), 2 | *Ad hoc* COPUOS (Legal Committee) 'Report of the Legal Committee' UN GAOR 14th Sessions UN Doc A/AC.98/2 (12 June 1959), 1.

[29]*Ad hoc* COPUOS (Legal Committee) 'Summary Record of the First Meeting' UN GAOR 14th Session UN Doc A/AC.98/C.2/SR.1 (30 June 1959), 5.

[30]*Ad hoc* COPUOS (Legal Committee) 'Summary Record of the First Meeting' UN GAOR 14th Session UN Doc A/AC.98/C.2/SR.1 (30 June 1959), 9.

[31]*Ad hoc* COPUOS 'Report of the Ad Hoc Committee on the Peaceful Uses of Outer Space' UN GAOR 14th Session UN Doc A/4141 (14 July 1959), 22.

[32]UNGA First Committee (14th Session) 'Summary Record of the 1079th Meeting' (11 December 1959) UN Doc A/C.1/SR.1079, par. 6.

interesting statement was made by India, which put forward that although COPUOS might initially focus on certain matters, the ultimate matter to be considered is the prohibition on the use of outer space for any military purposes whatsoever.[33] In addition, simultaneous progress would have to be made towards general disarmament.[34] Although India did not seem to question the mandate given to COPUOS, it foresaw the possibility that COPUOS could still deliberate on a prohibition of the use of outer space for military purposes.

These statements notwithstanding the draft resolution was adopted as Resolution 1472 (XIV), which states that the responsibility of COPUOS is as follows:

> (a) to review, as appropriate, the area of international co-operation, and to study practical and feasible means for giving effect to programmes in the ***peaceful (emphasis added)*** uses of outer space which could appropriately be undertaken under United Nations auspices, including, inter alia:
>
>> (i) Assistance for the continuation on a permanent basis of the research on outer space carried on within the framework of the International Geophysical Year;
>> (ii) Organization of the mutual exchange and dissemination of information on outer space research;
>> (iii) Encouragement of national research programmes for the study of outer space, and the rendering of all possible assistance and help towards their realization;
>
> (b) To study the nature of legal problems which may arise from the ***exploration (emphasis added)*** of outer space.[35]

In light of Resolution 1472 (XIV) and the deliberations between states prior to the adoption of the resolution, it can be unequivocally determined that the initial mandate of COPUOS is limited to the peaceful uses of outer space. This means that the disarmament of outer space is to be discussed outside of COPUOS. This mandate was reiterated in Resolution 1721 (XVI) E.[36]

In accordance with its mandate, COPUOS has been serving as a central platform for international cooperation in the field of outer space activities since its establishment. COPUOS has two subcommittees: the Scientific and Technical Subcommittee (STSC) and the Legal Subcommittee. These two subcommittees cooperate with each other and meet annually. Each year, COPUOS submits a report to the Fourth Committee of UNGA, which then adopts a resolution on 'International cooperation in the peaceful uses of outer space'. COPUOS aims at strengthening the international legal regime governing outer space and improving conditions for expanding international cooperation in the peaceful uses of outer space. As a focal point of international cooperation, the topics discussed in this forum include the following:

---

[33] UNGA First Committee (14th Session) 'Summary Record of the 1080th Meeting' (11 December 1959) UN Doc A/C.1/SR.1080, par. 12.

[34] UNGA First Committee (14th Session) 'Summary Record of the 1080th Meeting' (11 December 1959) UN Doc A/C.1/SR.1080, par. 12.

[35] UNGA Res 1472 (XIV) (12 December 1959).

[36] UNGA Res 1721 (XVI) (20 December 1961).

- maintaining outer space for peaceful purposes;
- safe operations in orbit;
- space debris;
- space weather;
- the threat from asteroids;
- the safe use of nuclear power in outer space;
- climate change;
- water management;
- global navigation satellite systems; and
- questions concerning space law and national space legislation.

The membership of COPUOS has continuously expanded, rising from 24 in 1959 to 87 member states in 2018.[37] This figure demonstrates the inclusivity of COPUOS, which has led to its success and indicates that the international community is paying increasing concern to the international governance of space activities.[38]

2.2 The Initial Mandate of the Conference on Disarmament (CD)

As stated, the discussion of space matters internationally began as a disarmament issue with Resolution 1148 (XII). The discussion of the peaceful uses of outer space was promptly transferred to a separate forum, but the disarmament of outer space still remained an issue to be discussed. The mandate to do this implicitly came to lie with the UN disarmament framework: the Ten-Nation Committee on Disarmament (Ten-Nation Committee), the Eighteen-Nation Committee on Disarmament (Eighteen-Nation Committee), the Conference of the Committee on Disarmament (CCD) and finally the Conference on Disarmament (CD). Although none of these committees had an explicit mandate to discuss the disarmament of outer space, they became the *de facto* forums for this matter.

2.2.1 Ten-Nation Committee

The issue of the disarmament of outer space had already cropped up in 1960 in discussions in the Ten-Nation Committee. France, for example, stated that the Committee's initial focus should be on nuclear disarmament as ballistic missiles and operational satellites strengthened the power of nuclear weapons to such a

---

[37]UNGA Res 1472 (XIV) (12 December 1959) | UNGA Res 72/77 (29 December 2017) UN Doc A/RES/72/77.
[38]Jessica West (ed), *Space Security Index 2017* (Ploughshares 2017) 127.

degree that conventional weapons and the strength of armed forces are secondary considerations.[39] The U.S. was even clearer, explicitly stating: 'We must take immediate action to prevent the extension of the arms race into outer space.'[40] Thus, a proposal that was put forward by Canada, France, Italy, the UK and the U.S. explicitly referred to outer space. This proposal emphasised the goal of ensuring the use of outer space for peaceful purposes only.[41] It was also more explicit about the disarmament of outer space, as stated by the UK:

> However, there are two points on which I have already touched to which I should like to return because, to my mind, they are very striking and of considerable importance to us.
>
> The first of these (...), provisions to ensure that weapons of mass destruction – particularly, of course, nuclear weapons – are not placed in orbit above the earth. This may seem to be legislating to the fantastic and to be more appropriate, perhaps, to a schoolboy's space fiction magazine than to serious international consideration, but I do ask my colleagues most earnestly not to treat the matter lightly.
>
> Today it is unquestionably possible to put very large weights, which could embrace nuclear weapons, into orbit round the world. (...) We must ensure that nuclear weapons are never put into orbit round the world by anyone. (...)
>
> It is for this reason that we have attached great importance to measures dealing with outer space and have introduced them at an early stage in our plan.[42]

In contrast, the USSR, and a number of like-minded states of the Warsaw Pact—Bulgaria, Czechoslovakia, Poland and Romania—specifically omitted mentioning outer space or the disarmament of outer space. Instead, their proposal was focused on 'general and complete disarmament', which has as its goal the full elimination of arms.[43] Therefore, although not explicitly mentioned, the disarmament of outer space was implicitly included in the proposal.

Eventually, however, the efforts in the Ten-Nation Committee collapsed, in part due to space related issues, as stated by the USSR:

> Unfortunately, instead of getting down to the drafting of a programme of disarmament in the Ten Nation Committee, the Western Powers have put forward proposals directed towards the establishment of control without disarmament; (...)
>
> It is significant that they place special emphasis on the establishment of control over military space rockets. It is easy to understand that the United States of America and its partners are

---

[39]UN Conference of the Ten Nation Committee on Disarmament, 'Final Verbatim Record of the First Meeting' (15 March 1960) UN Doc TNCD/PV.1, 16.

[40]UN Conference of the Ten Nation Committee on Disarmament, 'Final Verbatim Record of the First Meeting' (15 March 1960) UN Doc TNCD/PV.1, 35.

[41]UN Conference of the Ten Nation Committee on Disarmament, 'Final Verbatim Record of the Second Meeting' (16 March 1960) UN Doc TNCD/PV.2, 7.

[42]UN Conference of the Ten Nation Committee on Disarmament, 'Final Verbatim Record of the Second Meeting' (16 March 1960) UN Doc TNCD/PV.2, 12.

[43]UN Conference of the Ten Nation Committee on Disarmament, 'Final Verbatim Record of the Fifth Meeting' (21 March 1960) UN Doc TNCD/PV.5, 32–43 | UN Conference of the Ten Nation Committee on Disarmament, 'Final Verbatim Record of the Sixth Meeting' (22 March 1960) UN Doc TNCD/PV.6, 11–17 | UN Conference of the Ten Nation Committee on Disarmament, 'Final Verbatim Record of the Seventh Meeting' (23 March 1960) UN Doc TNCD/PV.7, 4–10.

hoping thereby to obtain unilateral military advantages for themselves and for the military blocs which they lead.[44]

This ultimately led to the USSR, Bulgaria, Czechoslovakia, Poland and Romania not attending the 48th meeting of the Ten-Nation Committee, which signalled the end of these discussions in the Committee.[45]

2.2.2 Eighteen-Nation Committee

Discussions on disarmament, and on disarmament in outer space, resumed in the Eighteen-Nation Committee. The deliberations included a proposal for a treaty banning nuclear weapon tests in the atmosphere, in outer space and underwater,[46] the proposal that eventually led to the PTBT.[47] The co-chairmen also explicitly recommended that the issue of 'Measures on the use of outer space for peaceful purposes only, together with appropriate control measures', be dealt with by the Committee.[48]

The inclusion of space matters in the discussion on general and complete disarmament also materialised in the draft treaties submitted by the USSR and the U.S. The USSR Draft Treaty referred to 'rocket devices' in the preamble[49] while also addressing space matters more explicitly. For example, article 5 of the Draft Treaty stated: 'All rockets capable of delivering nuclear weapons, of any calibre and range, whether strategic, operational or tactical (except for strictly limited numbers of rockets to be converted to peaceful uses).'[50] Article 5(4) of the Draft Treaty expanded on this by allowing the production and testing of rockets for the peaceful exploration of space provided there was supervision by the International Disarmament Organisation (an organisation to be established in accordance with the Draft Treaty).[51] This supervisory role of the International Disarmament Organisation over

---

[44]UN Conference of the Ten Nation Committee on Disarmament, 'Final Verbatim Record of the Forty-Seventh Meeting' (27 June 1960) UN Doc TNCD/PV.47, 4.

[45]UN Conference of the Ten Nation Committee on Disarmament, 'Final Verbatim Record of the Forty-Eight Meeting' (28 June 1960) UN Doc TNCD/PV.48, 5.

[46]UNGA 'Report to the United Nations of the Conference of the Eighteen-Nation Committee on Disarmament' (18 September 1962) UN Doc A/5200, 5.

[47]Treaty Banning Nuclear Tests in the Atmosphere, in Outer Space and Under Water (adopted 5 August 1963, entered into force 10 October 1963) 480 UNTS 43.

[48]UNGA 'Report to the United Nations of the Conference of the Eighteen-Nation Committee on Disarmament' (18 September 1962) UN Doc A/5200, 13.

[49]UN Disarmament Commission, 'Supplement for January 1961 to December 1962' (31 March 1963), 115.

[50]UN Disarmament Commission, 'Supplement for January 1961 to December 1962' (31 March 1963), 117.

[51]UN Disarmament Commission, 'Supplement for January 1961 to December 1962' (31 March 1963), 118.

peaceful uses of outer space (or at least rockets to be used for peaceful uses) was reiterated in article 15 of the Draft Treaty. First, article 15 stipulated that the launching of rockets and space devices shall be carried out exclusively for peaceful purposes.[52] Second, the International Disarmament Organisation would have control over the launch of rockets and space devices through inspection.[53] In addition, the USSR Draft Treaty included a provision that would later be included in the OST, namely the prohibition of placing into orbit or stationing in outer space of any device capable of delivering weapons of mass destruction.[54]

Likewise, the U.S. put forward an outline of basic provisions of a treaty on general and complete disarmament in a peaceful world. This outline also focused on the means of delivering weapons of mass destruction, which included rockets.[55] Moreover, the outline contained certain provisions specifically addressing outer space. The principles in this outline, like in its USSR counterpart, stipulated peaceful cooperation in outer space, notification and pre-launch inspection of space vehicles and missiles, limitation on the production of space vehicles and a prohibition on the placement of weapons of mass destruction in orbit.[56] The provisions in the USSR Draft Treaty and the principles in the U.S. Outline illustrate the interrelationship between space matters, whether military or non-military. Both the USSR Draft Treaty and the U.S. Outline would have a direct impact on the manner in which peaceful uses of outer space would and could be conducted.

Concurrent with the discussion on a treaty on general and complete disarmament, discussion also took place on a treaty banning nuclear weapon tests, including nuclear tests in outer space.[57] Inclusion of the topic of banning nuclear tests in outer space seems to have been uncontroversial because there was hardly any debate on whether or not to include outer space in the scope of the treaty. Rather, the debate focused on the level and manner of verification, as illustrated by a statement made by the U.S. in which the USSR proposal was criticised for not including control staff, control instruments or inspections.[58] The USSR had a different view, stating that it

---

[52] UN Disarmament Commission, 'Supplement for January 1961 to December 1962' (31 March 1963), 121.

[53] UN Disarmament Commission, 'Supplement for January 1961 to December 1962' (31 March 1963), 121–122.

[54] UN Disarmament Commission, 'Supplement for January 1961 to December 1962' (31 March 1963), 121.

[55] UN Disarmament Commission, 'Supplement for January 1961 to December 1962' (31 March 1963), 142.

[56] UN Disarmament Commission 'Supplement for January 1961 to December 1962' (31 March 1963), 146.

[57] UN Disarmament Commission 'Supplement for January 1961 to December 1962' (31 March 1963), 5, 11–14, 15–16.

[58] UN Disarmament Commission 'Supplement for January 1961 to December 1962' (31 March 1963), 18, 22–24 | Association for Diplomatic Studies, 'Moments in U.S. Diplomatic History: Negotiating the Limited Test Ban Treaty (LTBT)' <https://adst.org/2016/12/negotiating-limited-test-ban-treaty-ltbt/> accessed 23 June 2018 | Bureau of Arms Control, Verification and

was the policy of the Western Powers that prevented successful negotiations on a treaty banning nuclear tests and that international monitoring and supervision was sufficient.[59] The debate concentrated on the inclusion of the banning of underground nuclear tests. In contrast to the verification of nuclear tests in outer space, underwater or in the atmosphere, verification of underground nuclear tests was deemed to encounter serious technical difficulties.[60]

These points were heavily debated, but no real progress was made until the Cuban Missile Crisis.[61] The looming nuclear threat and close-call experienced during the crisis led to an increased effort to conclude a nuclear test ban treaty. The discussions resumed in June 1963 and led to an agreement between the USSR, the U.S. and the UK in July 1963.[62] Although these discussions took place within the context of the Eighteen-Nation Committee, the substantive negotiations took place between the aforementioned states as three of the four states that possessed nuclear weapons at that time (France was officially a member of the Eighteen-Nation Committee but never took its position in the Committee).[63]

This agreement resulted in the Partial Test Ban Treaty (PTBT), which quite simply stipulates in Article I as follows:

> Each of the Parties to this Treaty undertakes to prohibit, to prevent, and not to carry out any nuclear weapon test explosion, or any other nuclear explosion, at any place under its jurisdiction or control:
>
> (a) In the atmosphere; beyond its limits, including outer space; or under water, including territorial waters or high seas; (...).

In addition to the PTBT, 1963 saw the adoption of another space-related disarmament instrument. The Eighteen-Nation Committee put forward a draft resolution calling for a ban on the placement of nuclear weapons or other kinds of weapons of

---

Compliance, 'Treaty Banning Nuclear Weapon Tests in the Atmosphere, in Outer Space and Under Water' <https://www.state.gov/t/isn/4797.htm> accessed 23 June 2018.

[59]UN Disarmament Commission 'Supplement for January 1961 to December 1962' (31 March 1963), 44 | Bureau of Arms Control, Verification and Compliance, 'Treaty Banning Nuclear Weapon Tests in the Atmosphere, in Outer Space and Under Water' <https://www.state.gov/t/isn/4797.htm> accessed 23 June 2018.

[60]UN Disarmament Commission 'Supplement for January 1961 to December 1962' (31 March 1963), 43.

[61]Association for Diplomatic Studies, 'Moments in U.S. Diplomatic History: Negotiating the Limited Test Ban Treaty (LTBT)' <https://adst.org/2016/12/negotiating-limited-test-ban-treaty-ltbt/> accessed 23 June 2018 | Office of the Historian, 'The Cuban Missile Crisis, October 1962' <https://history.state.gov/milestones/1961-1968/cuban-missile-crisis> accessed 23 June 2018.

[62]UNGA 'Report of the Conference of the Eighteen-Nation Committee on Disarmament' (5 September 1963) UN Doc A/5488-DC/208.

[63]Association for Diplomatic Studies, 'Moments in U.S. Diplomatic History: Negotiating the Limited Test Ban Treaty (LTBT)' <https://adst.org/2016/12/negotiating-limited-test-ban-treaty-ltbt/> accessed 23 June 2018.

mass destruction in orbit in outer space.[64] This draft resolution was a collaborative effort between the USSR and the U.S.[65] Resolution 1884 (XVIII) was adopted without a vote, but both the USSR and the U.S. made statements about the resolution. The USSR stated that the prevention of an arms race in outer space was part of its policy.[66] A similar statement was made by the U.S., namely that the newly explored environment of outer space should be kept free of nuclear weapons and other weapons of mass destruction.[67]

Mexico made two interesting points illustrating its vision of the use of outer space. First, 'It would be useless for nations to endeavour to disarm on Earth if they were to arm in outer space, and it would be vain to try to denuclearise various zones of the Earth if the nuclearization of outer space were not prevented'.[68] Second, Mexico stated: 'Let us express, by means of our vote, our unanimous hope and determination that the space ships of today, tomorrow and the future will always, and solely, be messengers of peace.'[69] Both these statements indicate a strong desire to prohibit any kind of military use of outer space.

The result was Resolution 1884 (XVIII), which documents the will of the USSR and the U.S. not to station nuclear weapons or other weapons of mass destruction in orbit and calls on all states not to place into orbit, install on celestial bodies or station in outer space such weapons.[70] Eventually, this resolution was superseded by the adoption of the OST, which in Article IV stipulates the same, but now legally binding, prohibition.

After the adoption of Resolution 1884 (XVIII), the Eighteen-Nation Committee continued its discussion on disarmament, however, without a particular focus on

---

[64]UNGA 'Question of General and Complete Disarmament: Report of the Conference of the Eighteen-Nation Committee on Disarmament' UN GAOR 18th Session UN Doc A/5571 (17 October 1963), 1.

[65]UNGA 'Question of General and Complete Disarmament: Report of the Conference of the Eighteen-Nation Committee on Disarmament' UN GAOR 18th Session UN Doc A/5571 (17 October 1963), 1.

[66]UNGA 'Question of General and Complete Disarmament: Report of the Conference of the Eighteen-Nation Committee on Disarmament' UN GAOR 18th Session UN Doc A/5571 (17 October 1963), 1.

[67]UNGA 'Question of General and Complete Disarmament: Report of the Conference of the Eighteen-Nation Committee on Disarmament' UN GAOR 18th Session UN Doc A/5571 (17 October 1963), 2.

[68]UNGA 'Question of General and Complete Disarmament: Report of the Conference of the Eighteen-Nation Committee on Disarmament' UN GAOR 18th Session UN Doc A/5571 (17 October 1963), 2.

[69]UNGA 'Question of General and Complete Disarmament: Report of the Conference of the Eighteen-Nation Committee on Disarmament' UN GAOR 18th Session UN Doc A/5571 (17 October 1963), 2.

[70]UNGA Res 1884 (XVIII) (17 October 1963).

outer space.[71] Instead, the focus was on negotiating a treaty on general and complete disarmament and a treaty to prevent the spread of nuclear weapons,[72] which became the Non-Proliferation Treaty.[73]

2.2.3 Conference of the Committee on Disarmament

From 1969 onwards, the Eighteen-Nation Committee once more expanded. In addition to the 18 states that were already participating, Japan, Mongolia, Argentina, Hungary, Morocco, the Netherlands, Pakistan and Yugoslavia joined to create a better 'geographic and political balance', leading to the Committee being renamed the 'Committee on Disarmament' and the Conference the 'Conference of the Committee on Disarmament'.[74] Once more, space matters were not explicitly specified in the mandate. However, the provisional agenda that was adopted by the CCD included 'other collateral measures,'[75] examples of which were the prevention of an arms race on the seabed or similar measures.[76] In light of the previous discussion of space matters in the Ten-Nation Committee and Eighteen-Nation Committee, and the example given in the provisional agenda, space matters were implicitly included in the mandate of the CCD. This is further reinforced by the express mention of both the PTBT and the OST as examples of then-recent disarmament negotiations.[77]

Nevertheless, space matters were sparsely discussed in the CCD. Instead, the CCD focused on general and complete disarmament and a comprehensive test ban treaty.[78] In fact, negotiations concentrated on a treaty dealing with the prevention of an arms race on the seabed and the ocean floor, which resulted in the Seabed Arms Control Treaty.[79]

---

[71] UNGA 'Report of the Conference of the Eighteen-Nation Committee on Disarmament (21 January - 17 September 1964)' (22 September 1964) UN Doc A/5731-DC/209 | UNGA 'Report of the Eighteen-Nation Committee on Disarmament' (22 September 1965) UN Doc A/5986-DC/227.

[72] UN Disarmament Commission 'Supplement for 1966' (1967) UN Doc DC/228 | UN Disarmament Commission 'Supplement for 1967 and 1968' (1969) UN Doc DC/230 & DC/231.

[73] Treaty on the Non-Proliferation on Nuclear Weapons (adopted 1 July 1968, entered into force 5 March 1970) 729 UNTS 161 [hereinafter: NPT].

[74] UN Disarmament Commission 'Supplement for 1969' (1971) UN Doc DC/232, 2.

[75] UN Disarmament Commission 'Supplement for 1969' (1971) UN Doc DC/232, 2.

[76] UN Disarmament Commission 'Supplement for 1969' (1971) UN Doc DC/232, 2.

[77] UN Disarmament Commission 'Supplement for 1969' (1971) UN Doc DC/232, 2–3.

[78] UN Disarmament Commission 'Supplement for 1970' (1971) UN Doc DC/233 | UN Disarmament Commission 'Supplement for 1971' (1973) UN Doc DC/234 | UN Disarmament Commission 'Supplement for 1972' (1974) UN Doc DC/235.

[79] Treaty on the Prohibition of the Emplacement of Nuclear Weapons and Other Weapons of Mass Destruction on the Sea-Bed and the Ocean Floor and in the Subsoil Thereof (*adopted* 11 February 1971, *entered into force* 18 May 1972) 955 UNTS 115.

No similar negotiations were necessary for outer space because of the adoption of the OST, which already contained a similar provision in Article IV. However, this does not mean that the CCD did not consider space matters, at least as a part of overarching issues. For example, space matters were discussed during the negotiations on a treaty on the limitation of anti-ballistic missile systems, which included an obligation not to develop, test or deploy such systems that are space based.[80] In addition, outer space was included in the negotiations on meteorological warfare as one of the environments that shall not be modified for military or any other hostile use.[81] Ultimately, outer space was included under Article II of the Environmental Modification Convention.[82] No further discussion on outer space took place, except for a number of references in the overarching discussions on disarmament.[83]

2.2.4 *The Conference on Disarmament*

Only when the Committee on Disarmament, which was renamed the 'Conference on Disarmament' in 1984, was re-established in 1979 was an explicit mandate stipulated. In general, the programme established in the first special session of the UNGA devoted to disarmament, as laid down in Resolution S-10/2, has as its objective to 'halt the arms race in all its aspects, to open a process of genuine disarmament on an internationally agreed basis and to increase international confidence and relaxation of international tension'.[84] The long-term objective is then to 'achieve general and complete disarmament under effective international control, to avert the danger of war and to create conditions for a just and stable international peace and security and the full realization of the new international economic order'.[85]

In addition, Resolution S-10/2 recommended establishing a new negotiation body, stating: 'The Assembly is deeply aware of the continuing requirement for a single multilateral disarmament negotiating forum of limited size taking decisions on

---

[80] UNGA 'Report of the Conference of the Committee on Disarmament' UN GAOR 28th Session Supp No 31 UN Doc A/9141 (1975), 42.

[81] UNGA 'Report of the Conference of the Committee on Disarmament Volume I' UN GAOR 31st Session Supp No 27 UN Doc A/31/27 (1976), 87.

[82] Convention on the Prohibition of Military or any other Hostile Use of Environmental Modification Techniques (*adopted* 10 December 1976, *entered into force* 5 October 1978) 1108 UNTS 151.

[83] UNGA 'Report of the Conference of the Committee on Disarmament Volume I' UN GAOR 32nd Session Supp No 27 UN Doc A/32/27 (1977), 12 | UNGA 'Report of the Conference of the Committee on Disarmament Volume II' UN GAOR 32nd Session Supp No 27 UN Doc A/32/27 (1977), 18.

[84] UNGA 'Report of the Disarmament Commission' UN GAOR 34th Session Supp No 42 UN Doc A/34/42 (25 June 1979), par. 8.

[85] UNGA 'Report of the Disarmament Commission' UN GAOR 34th Session Supp No 42 UN Doc A/34/42 (25 June 1979), par. 9.

the basis of consensus.'[86] The general mandate of the negotiating forum, that is the CD, is thus to negotiate multilateral disarmament decisions on the basis of consensus; within this general mandate, the CD is free to adopt its own agenda taking into account the recommendations made to it by the UNGA and the proposals presented by the members of the Committee.[87]

The prevention of an arms race in outer space specifically became part of the CD agenda because Resolution S-10/2 recommended that further measures should be taken and appropriate international negotiations should be held in accordance with the OST to prevent such an arms race.[88] The recommendation to include the prevention of an arms race in outer space on the CD agenda was subsequently adopted and implemented by the Disarmament Commission.[89] The mandate to discuss space-related disarmament issues was affirmed by the UNGA in Resolution 36/97-C, which stated that the UNGA

> 1. Considers that further effective measures *to prevent an arms race in outer space (emphasis added)* should be adopted by the international community;
>
> (...)
>
> 3. Requests the Committee on Disarmament to consider, as from the beginning in 1982, the question of negotiating effective and verifiable agreements aimed at *preventing an arms race in outer space (emphasis added)*, taking into account all existing and future proposals designed to meet this objective;
>
> 4. Requests the Committee on Disarmament to consider as a matter of priority the question of negotiating an effective and verifiable agreement to *prohibit anti-satellite systems (emphasis added)*, as an important step towards the fulfilment of the objectives set out in paragraph 3 above; (...).[90]

In addition, in its Resolution 36/99, the General Assembly specifically requested the CD to deliberate on the conclusion of a *treaty on the prohibition of the stationing of weapons of any kind in outer space* as follows:

> 1. Considers it necessary to take effective steps, by concluding an appropriate international treaty, to prevent the spread of the arms race to outer space;
>
> 2. Requests the Committee on Disarmament to embark on negotiations with a view to achieving agreement on the text of such a treaty.[91]

---

[86] UNGA Res S-10/2 'Final Document of the Tenth Special Session of the General Assembly' (28 June 1978) UN Doc A/RES/S-10/2, par. 120.

[87] UNGA Res S-10/2 'Final Document of the Tenth Special Session of the General Assembly' (28 June 1978) UN Doc A/RES/S-10/2, par. 120(e).

[88] UNGA Res S-10/2 'Final Document of the Tenth Special Session of the General Assembly' (28 June 1978) UN Doc A/RES/S-10/2, par. 80.

[89] UNGA 'Report of the Disarmament Commission' UN GAOR 36th Session Supp 42 UN Doc A/36/42 (2 July 1981), par. 19.

[90] UNGA Res 36/97-C (9 December 1981) UN Doc A/RES/36/97-C.

[91] UNGA Res 36/99 (9 December 1981) UN Doc A/RES/36/99.

In light of the aforementioned, the CD thus received a full mandate to discuss the disarmament of outer space internationally.

The CD was established as a single multilateral disarmament negotiating a forum of the international community in 1979. It addresses the disarmament of outer space through initiatives. With 65 current member states, the CD works by consensus under a rotating presidency.[92] Its permanent agenda is known as the Decalogue and includes the following issues:

- nuclear weapons in all aspects;
- other weapons of mass destruction;
- conventional weapons;
- reduction of military budgets;
- reduction of armed forces;
- disarmament and development;
- disarmament and international security;
- collateral measures;
- confidence-building measures;
- effective verification methods in relation to appropriate disarmament measures, acceptable to all parties; and
- comprehensive programme of disarmament leading to general and complete disarmament under effective international control.

Similar to the working mechanism of COPUOS, the CD is requested to report to the UNGA annually, or more frequently when it finds appropriate. As a negotiating forum, the CD functions differently from the Disarmament Commission, for the latter is a deliberative body. The CD and the Disarmament Commission constitute the existing machinery in the field of disarmament to help the UN fulfil its role.

2.3 The Discussion of Space Matters Internationally Based on the Initial Mandates of COPUOS and the CD

In light of the aforementioned, the conclusion might be drawn that COPUOS and the CD have different but complementing mandates to discuss space matters internationally; COPUOS is mandated to deliberate on the peaceful uses of outer space, while the CD encompasses all disarmament issues and therefore also has the mandate to discuss the disarmament of outer space. This then suggests that the discussion of space matters internationally is settled.

This is, however, not the case. The mandates bestowed upon COPUOS and the CD are not without issues. On the one hand, the mandates presume that space matters can be distinguished into separate categories, which can be discussed separately. This perspective, however, is short-sighted and does not consider the reality of space

---

[92] Jessica West (ed), *Space Security Index 2017* (Ploughshares 2017) 129.

activities. Space technologies are intrinsically dual use in nature.[93] Furthermore, it is apparent that the use of outer space for exclusively peaceful uses can only be achieved after an arms race in outer space is prevented. The interrelation between the discussion of peaceful uses of outer space and the disarmament of outer space also becomes apparent in the deliberations undertaken in the forums. It was the CD that discussed the obligation in UNGA Resolution 1884 (XVIII) to refrain from placing into orbit nuclear weapons and other kinds of weapons of mass destruction. After the adoption of this resolution, COPUOS discussed this issue further, which eventually led to the inclusion of a similar obligation in Article IV OST.[94] The discussion, and subsequent adoption, of an obligation in one of the forums leading to the further discussion of this obligation in the other forum illustrates that space matters are interrelated and cannot be discussed separately as the given mandates indicate. This is further illustrated by proposals submitted by both the USSR and the U.S. in the discussions in the CD on a treaty on general and complete disarmament. Although these drafts never came to fruition, both contained provisions that would limit the number of rockets that could be produced and used, even for peaceful purposes.[95]

On the other hand, there is a question as to whether the mandates given to the forums cover all space matters and whether there is a void in the discussion of space matters internationally. COPUOS is limited to discussing peaceful uses of outer space, with uncertainty as to what uses are actually considered peaceful. In contrast, the CD does not discuss all military uses of outer space but is limited to disarmament. Therefore, depending on the interpretation of the term 'peaceful', two conclusions could be drawn. First, 'peaceful' could be interpreted as 'non-military'. This would mean that the discussion of military uses of outer space would fall outside the mandate of COPUOS. However, because the mandate of the CD limits itself to disarmament, the 'non-arms military' uses of outer space,[96] such as remote sensing, telecommunications or GPS for military purposes,[97] would then fall outside the

---

[93] Anél Ferreira-Snyman, 'Selected Legal Challenges Relating to the Military Use of Outer Space, With Specific Reference to Article VI of the Outer Space Treaty' (2015) 18 Potchefstroom Electronic Law Journal 488, 495–496.

[94] UN, 'Comprehensive Study of the Question of Nuclear-Weapon-Free Zones in All Its Aspects' (1976) UN Doc A/10027/Add.1, 12.

[95] UN Disarmament Commission 'Official Records: Supplement for January 1961 to December 1962' (31 March 1963), 117 & 142.

[96] Throughout this book reference will be made to the 'non-arms military' use of outer space. This term is not based on an official term used in COPUOS or the CD. Rather, it is used in this publication to make a distinction between the military uses of outer space that make use of arms, which definitely fall within the scope of the disarmament of outer space, and military uses of outer space that do not make use of arms, which depending on the interpretation of the mandates of COPUOS and the CD could fall within the scope of the disarmament of outer space or within the scope of the peaceful use of outer space. The use of the term 'non-arms military' aims to contribute to the clarity of the research.

[97] David A. Koplow, 'The Fault Is Not in Our Stars: Avoiding an Arms Race in Outer Space' (2018) 59 Harvard International Law Journal 331, 335–336 | Anél Ferreira-Snyman, 'Selected Legal

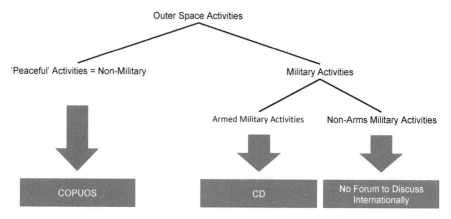

Fig. 2.1 If peaceful is interpreted as non-military

mandate of both forums and would not have a forum in which they are to be discussed internationally, as illustrated in Fig. 2.1.

The second option is that the term 'peaceful' is interpreted as 'non-aggressive'. In general, the 'non-arms military' uses of outer space would then fall within the mandate of COPUOS, as illustrated in Fig. 2.2.

Even in this case, however, a lacuna still exists as it could be argued that there are uses of outer space that are aggressive yet not related to arms or disarmament. Indeed, there are few such uses, and it also depends on what definitions are used for the terms 'arms', 'aggressive', and 'disarmament'. Examples of uses that are not necessarily arms but can be aggressive are kinetic interceptors and cyberattacks.[98] Therefore, such uses could still be excluded from international deliberations.

Therefore, the question is whether the term 'peaceful' is equal to 'non-military', or does it mean 'non-aggressive' and thus includes 'non-arms military' uses of outer space. Because the term 'peaceful' has never been defined in any of the UN instruments dealing with space,[99] or indeed in any legal instruments at all, what the term encompasses has long been discussed.[100] In the early stages of the development of space law, the discussion could be divided between the U.S. position and the USSR position. The position put forward by the USSR was that of the 'non-

---

Challenges Relating to the Military Use of Outer Space, With Specific Reference to Article VI of the Outer Space Treaty' (2015) 18 Potchefstroom Electronic Law Journal 488, 495–496.

[98]David A. Koplow, 'The Fault Is Not in Our Stars: Avoiding an Arms Race in Outer Space' (2018) 59 Harvard International Law Journal 331, 339.

[99]Francis Grimal & Jae Sundaram, 'The Incremental Militarization of Outer Space: A Threshold Analysis' (2018) 17 Chinese Journal of International Law 45, 51 | Anél Ferreira-Snyman, 'Selected Legal Challenges Relating to the Military Use of Outer Space, With Specific Reference to Article VI of the Outer Space Treaty' (2015) 18 Potchefstroom Electronic Law Journal 488, 496.

[100]Fabio Tronchetti, 'Legal Aspects of the Military Uses of Outer Space' in Frans von der Dunk & Fabio Tronchetti (eds), *Handbook of Space Law* (Edward Elgar Publishing 2015) 331.

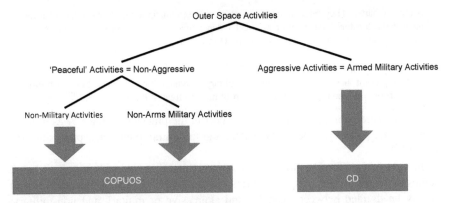

Fig. 2.2 If peaceful is interpreted as non-aggressive

military' interpretation of the term 'peaceful'.[101] This approach argues that the use of outer space for exclusively peaceful purposes indicates that all military uses of outer space are prohibited.[102] This interpretation is based on the use of the term 'peaceful' in the Antarctic Treaty,[103] which also further expands on the term, stating in Article I:

> There shall be prohibited, inter alia, any measures of a military nature, such as the establishment of military bases and fortifications, the carrying out of military maneuvers, as well as the testing of any type of weapons.[104]

This approach was initially also supported by the U.S., but the U.S. position quickly shifted towards the 'non-aggressive' interpretation.[105] This is best exemplified in a released Central Intelligence Agency (CIA) memorandum that states:

> In the context of outer space activities, the United States seeks to counter any efforts to equate the distinction between 'peaceful' and 'aggressive' with that between 'civilian' and 'military'. Though the United States does not view further definition of the terms 'peaceful uses' or 'peaceful purposes' as necessary or appropriate at this time, our view is that these terms clearly refer to activities consistent with international law, including the United

---

[101] Fabio Tronchetti, 'Legal Aspects of the Military Uses of Outer Space' in Frans von der Dunk & Fabio Tronchetti (eds), *Handbook of Space Law* (Edward Elgar Publishing 2015) 339.

[102] Fabio Tronchetti, 'Legal Aspects of the Military Uses of Outer Space' in Frans von der Dunk & Fabio Tronchetti (eds), *Handbook of Space Law* (Edward Elgar Publishing 2015) 339 | Anél Ferreira-Snyman, 'Selected Legal Challenges Relating to the Military Use of Outer Space, With Specific Reference to Article VI of the Outer Space Treaty' (2015) 18 Potchefstroom Electronic Law Journal 488, 497.

[103] Anél Ferreira-Snyman, 'Selected Legal Challenges Relating to the Military Use of Outer Space, With Specific Reference to Article VI of the Outer Space Treaty' (2015) 18 Potchefstroom Electronic Law Journal 488, 496.

[104] The Antarctic Treaty (adopted 1 December 1959, entered into force 23 June 1961) 402 UNTS 71.

[105] Francis Grimal & Jae Sundaram, 'The Incremental Militarization of Outer Space: A Threshold Analysis' (2018) 17 Chinese Journal of International Law 45, 51.

Nations Charter. They refer to activities which do not constitute the threat or use of force against the territorial integrity or political independence of any state, and are not in any matter inconsistent with the purposes of the United Nations.

(...)

Thus at present there are no restraints on military activities in peace (**most likely a typing error in the document, read: space**) short of use or threat of the use of force.[106]

The interpretation has found large support in the practice of states[107] as military uses of outer space have been conducted, and more importantly accepted, since the 1950s.[108]

In light of the aforementioned, it is apparent that there are uncertainties when it comes to the discussion of space matters internationally. First, space matters cannot simply be divided between peaceful and aggressive or military and non-military. They are inherently intertwined. Therefore, these matters cannot be adequately discussed in separate forums. At the very least, there should be a continuous collaboration between the forums to adequately discuss space matters. Second, the mandates given to COPUOS and the CD create a potential void in the discussion of space matters internationally. Depending on the favoured definitions of the terms 'peaceful', 'arms', 'disarmament' and 'aggressive', the scope of the given mandates might vary. This can lead to certain 'non-arms military' uses of outer space going without a forum in which they are to be discussed internationally. State practice strongly indicates that 'peaceful' is interpreted as 'non-aggressive'. In this case, the void is non-existent or, at the very least, very narrow. However, this should also mean that 'non-arms military' uses of outer space are actively discussed in COPUOS as they are 'non-aggressive' and thus a peaceful use of outer space. A cursory glance at the actual deliberations in COPUOS indicates that this is not the case. Therefore, the next chapter will go through the deliberations in COPUOS and the CD to determine the mandate of the forums as interpreted in practice and to determine whether the outlined void exists.

---

[106] CIA, Definition of Peaceful Uses of Outer Space, Document Type: Crest, Document Number: CIA-RDP66R00638R000100160004-2 Approved for release 28 August 2001 <https://www.cia.gov/library/readingroom/docs/CIA-RDP66R00638R000100160004-2.pdf> accessed 29 June 2018, 1–2.

[107] Anél Ferreira-Snyman, 'Selected Legal Challenges Relating to the Military Use of Outer Space, With Specific Reference to Article VI of the Outer Space Treaty' (2015) 18 Potchefstroom Electronic Law Journal 488, 497.

[108] Fabio Tronchetti, 'Legal Aspects of the Military Uses of Outer Space' in Frans von der Dunk & Fabio Tronchetti (eds), *Handbook of Space Law* (Edward Elgar Publishing 2015) 339–340 | Steven Freeland, 'In Heaven as on Earth? The International Legal Regulation of the Military Use of Outer Space' (2011) 8 U.S.-China Law Review 272, 277 | Francis Grimal & Jae Sundaram, 'The Incremental Militarization of Outer Space: A Threshold Analysis' (2018) 17 Chinese Journal of International Law 45, 46 & 54–55 | Anél Ferreira-Snyman, 'Selected Legal Challenges Relating to the Military Use of Outer Space, With Specific Reference to Article VI of the Outer Space Treaty' (2015) 18 Potchefstroom Electronic Law Journal 488, 488–489.

Chapter 3
The Development of the Mandates of the Committee on the Peaceful Uses of Outer Space (COPUOS) and the Conference on Disarmament (CD) and the Collaboration Between the Forums

Annette Froehlich, Vincent Seffinga, and Ruiyan Qiu

Abstract The respective mandates initially given to the Committee on the Peaceful Uses of Outer Space (COPUOS) and the Conference on Disarmament (CD) leave a possible gap in the discussion of space matters internationally, especially since the discussion and deliberation of space matters internationally has evolved since the establishment of the forums. The purpose of this chapter is to focus on this development. In particular, it focuses on the evolution of the mandates of the two forums from their respective establishment until the present and the relations and collaboration between the forums. To frame the evolution of the discussion of space matters internationally, this chapter first gives a basic historical background overview. This overview is not meant to be exhaustive and all-encompassing but is intended to give a better frame of reference for the analysis that will follow.

The overview discusses the USSR–U.S. cold war and space race, European integration and space cooperation and the position of the People's Republic of China within the UN. Thereafter, it examines the development of the mandate of COPUOS. This examination shows that just after the establishment of COPUOS, three interpretations of the mandate existed: first, that all military uses of outer space should be discussed in the UN disarmament framework and that non-military uses of outer space should be discussed in COPUOS; second, that 'arms military' uses of

A. Froehlich (✉) · V. Seffinga
European Space Policy Institute, Vienna, Austria
e-mail: annette.froehlich@espi.or.at

R. Qiu
International Institute of Air and Space Law, Leiden University, Leiden, Netherlands

outer space should be discussed in the disarmament framework but that 'non-arms military uses' and non-military uses should be discussed in COPUOS; and third, that pure disarmament issues should be discussed in the disarmament framework but that, in addition to the 'non-arms military uses' and non-military uses, COPUOS is also mandated to discuss the limitation of the use of outer space to ensure that it is used exclusively for peaceful purposes.

During the deliberations on the five UN Space Treaties, these interpretations seemed to change, with the last interpretation prevailing. This is illustrated by the inclusion of Article IV OST and statements made by states. Following the adoption of Resolution S-10/2, however, the diverging interpretations reappeared. In addition, a fourth interpretation arose that called for close-knit cooperation between COPUOS and the CD. These four interpretations are still the status quo on the mandate of COPUOS. This chapter then examines the development of the mandate of the CD with respect to the discussion of outer space and shows that the CD, just like COPUOS, discusses 'non-arms military' uses of outer space and even exceeds its mandate by discussing certain peaceful uses of outer space. Lastly, the chapter reflects on the interaction and collaboration between the two forums, concluding that such interaction and collaboration is nearly non-existent.

The previous chapter set out the mandates of the Committee on the Peaceful Uses of Outer Space (COPUOS) and the Conference on Disarmament (CD) at the time of their respective establishment and has shown that there is a possible void in the discussion of space matters internationally, depending on the interpretation of the term 'peaceful' in the mandate of COPUOS. Since the establishment of the forums, the discussion and deliberation of space matters internationally have evolved. The purpose of this chapter is to focus on this development. In particular, it will focus on the evolution of the mandate of the two forums from their respective establishment until the present and the relation and collaboration between the forums.

To frame the evolution of the discussion of space matters internationally, this chapter shall first give a basic historical background overview (Sect. 3.1). This overview is not meant to be exhaustive and all-encompassing but meant to give a better frame of reference for the analysis that will follow. The overview will discuss the USSR–U.S. cold war and space race, European integration and space cooperation, and the position of the People's Republic of China (PRC) within the United Nations (UN). Thereafter, this chapter shall examine the development of the mandate of COPUOS (Sect. 3.2) and the mandate of the CD with respect to the discussion of outer space (Sect. 3.3). The research will examine a 50-year period. Due to the large number of documents produced in this period, it will not be feasible to examine them all. Instead, the research will focus on the major resolutions and instruments that have been discussed by the forums. Therefore, with respect to COPUOS, the research will examine the Declaration of Legal Principles Governing the Activities of States in the

3 The Development of the Mandates of the Committee on the Peaceful Uses... 31

Exploration and Use of Outer Space,[1] the Five UN Space Treaties[2] and the other major resolutions adopted.[3] It will also deal with relevant deliberations and initiatives that have not (yet) resulted in an instrument, such as the deliberation on the long-term sustainability of outer space activities (LTS). In contrast, fewer major instruments were adopted by the CD in the aforementioned time period. Thus, the resolutions, reports and other documents relevant to the discussion of the prevention of an arms race in outer space in the CD will be considered, as well as initiatives such as the Treaty on the Prevention of the Placement of Weapons in Outer Space (PPWT). Lastly, the chapter will discuss the interaction and collaboration between the two forums and the evolution of the mandates of COPUOS and the CD to discuss such issues (Sect. 3.4).

3.1 Historical Background

Discussion of the further development of the mandate of both COPUOS and the CD cannot be detached from the overall historical political-diplomatic and socio-economic developments. This paragraph will therefore examine the most important of these developments to frame the subsequent analysis of the development of the mandate of COPUOS and the CD. It will not attempt an exhaustive historical recount but merely a general overview to give a more complete understanding of the

---

[1] UNGA Res 1962 (XVIII) 'Declaration of Legal Principles Governing the Activities of States in the Exploration and Use of Outer Space' (13 December 1963) UN Doc A/RES/18/1962.

[2] Treaty on Principles Governing the Activities of States in the Exploration and Use of Outer Space, including the Moon and Other Celestial Bodies (adopted 19 December 1966, entered into force 10 October 1967) 610 UNTS 205 | Convention on the International Liability for Damage Caused by Space Objects (adopted 29 March 1972, entered into force 1 September 1972) 961 UNTS 187 | Convention on Registration of Objects Launched into Outer Space (adopted 12 November 1974, entered into force 15 September 1976) 1023 UNTS 15 | Agreement on the Rescue of Astronauts, the Return of Astronauts and the Return of Objects Launched Into Outer Space (adopted 22 April 1968, entered into force 3 December 1968) 672 UNTS 119 | Agreement Governing the Activities of States on the Moon and Other Celestial Bodies (adopted 5 December 1979, entered into force 11 July 1984) 1363 UNTS 21.

[3] UNGA Res 37/92 'The Principles Governing the Use by States of Artificial Earth Satellites for International Direct Television Broadcasting' (10 December 1982) UN Doc A/RES/37/92 | UNGA Res 41/65 'The Principles Relating to Remote Sensing of the Earth from Outer Space' (3 December 1986) UN Doc A/RES/41/65 | UNGA Res 47/68 'The Principles Relevant to the Use of Nuclear Power Sources in Outer Space' (14 December 1992) UN Doc A/RES/47/68 | UNGA Res 51/122 'The Declaration on International Cooperation in the Exploration and use of Outer Space for the Benefit and in the Interest of All States, Taking into Particular Account the Needs of Developing Countries' (13 December 1996) UN Doc A/RES/41/122 | UNGA Res 62/101 'Recommendations on Enhancing the Practice of States and International Intergovernmental Organizations in Registering Space Objects' (2007) UN Doc A/RES/62/101 | UNCOPUOS, 'Space Debris Mitigation Guidelines of the Committee on the Peaceful Uses of Outer Space' (2007) UN Doc A/62/20, endorsed by UNGA Res 62/217 (22 December 2007) UN Doc A/RES/62/217.

subsequent discussion. Specific topics include the USSR–U.S. Cold War and the space race, European integration and space cooperation, and the position of the People's Republic of China in the UN. These topics illustrate the political-diplomatic and socio-economic developments of the major spacefaring states from the establishment of COPUOS and the CD to the present day.

3.1.1 The USSR–U.S. Cold War and Space Race

Much of the discussion of international matters after the Second World War has been dominated by the tension and conflict between the USSR and the U.S., which affected nearly every state in the world.[4] In essence, the Cold War between the USSR and the U.S. (or the 'East' and the 'West') was both a struggle for power (resulting in an arms race between the two powers and proxy wars in Korea, Vietnam, Afghanistan, etc.) and an ideological struggle between the communist USSR and the capitalist U.S.[5] Although the USSR and the U.S. were allies during the Second World War, their opposing ideologies quickly turned them into adversaries in the closing stages of the war.[6] The defeat of Germany and Japan led to a power vacuum, which was then filled by the USSR and the U.S., with the U.S. in a stronger position.[7]

Although both the U.S. and the USSR quickly demobilised,[8] Soviet pressure on Turkey and the Greek Civil War prompted President Truman to state: 'It must be the policy of the United States to support free peoples who are resisting attempted subjugation by armed minorities or by outside pressures (...) we must assist free peoples to work out their own destinies in their own way.'[9] This statement ushered in an era of U.S. involvement in global affairs, specifically against communist governments and regimes, for example in Asia and Latin America. At the same time, the USSR sought to increase and solidify its own security: domestically, in its satellite states, and globally.[10] Both superpowers interfered and intervened in many conflicts around the globe, the most well-known being the Korean War, the Vietnam War and the Afghanistan War.

This tension could be seen throughout discussions on practically all security-related international matters. In the previous chapter, the example was given of the

---

[4]Melvyn Leffler and David Painter (eds), *Origins of the Cold War: An International History* (2nd Edition Routledge 2005) 1.

[5]Jeremy Black, *The Cold War: A Military History* (Bloomsbury 2015) 41, 51–57.

[6]Jeremy Black, *The Cold War: A Military History* (Bloomsbury 2015) 37–38.

[7]Melvyn Leffler and David Painter (eds), *Origins of the Cold War: An International History* (2nd Edition Routledge 2005) 3.

[8]Melvyn Leffler and David Painter (eds), *Origins of the Cold War: An International History* (2nd Edition Routledge 2005) 3–4.

[9]John Lewis Gaddis, *The Cold War: A New History* (The Penguin Press 2005) 73 | John Lamberton Harper, *The Cold War* (Oxford University Press 2011) 64.

[10]Jeremy Black, *The Cold War: A Military History* (Bloomsbury 2015) 49–57.

Ten-Nation Committee, which collapsed due to the U.S.–USSR power struggle. The USSR saw the U.S. and Western European position on disarmament as a way for the 'Western Bloc' to gain a unilateral military advantage over the USSR and its allies.[11] This ultimately led to the USSR, Bulgaria, Czechoslovakia, Poland and Romania not attending the 48th meeting of the Ten-Nation Committee, which signalled the end of the discussions in the Committee.[12]

This tension continued until the dissolution of the USSR and realistically still continues into the present day. It should be noted that the tension between the USSR and the U.S. varied throughout the Cold War, with moments of extreme tension, such as the Cuban Missile Crisis, and moments of reduced tension or *détente*,[13] for example during Khrushchev's Thaw.[14] Correspondingly, the level of tension between the two superpowers influenced and guided the discussion of international matters. In general, both states held to their ideological position at moments of extreme tension, while reduced tension allowed for compromises. In contrast, times of extreme tension could also serve as such a strong warning that compromises were made to avoid worst case scenarios. For example, the Cuban Missile Crisis and the fact that the world had been very close to a nuclear war led to progress in the negotiation and subsequent conclusion of the PTBT.[15]

This tension naturally also influenced the development of space technology. Both the first satellite launched into orbit by the USSR, Sputnik 1 in 1957, and the U.S., Explorer 1 in 1958, were launched as part of the International Geophysical Year, a remarkable feat of scientific cooperation during the Cold War.[16] However, the fact that both the USSR and the U.S. now had independent access to outer space resulted in a competition for supremacy over this new domain.[17] This competition, however, focused heavily on the civilian aspect of space technology.[18] That is not to say that neither the U.S. nor the USSR started developing military space technology (the first

---

[11]UN Conference of the Ten Nation Committee on Disarmament 'Final Verbatim Record of the Forty-Seventh Meeting' (27 June 1960) UN Doc TNCD/PV.47, 4.

[12]UN Conference of the Ten Nation Committee on Disarmament 'Final Verbatim Record of the Forty-Eight Meeting' (28 June 1960) UN Doc TNCD/PV.48, 5.

[13]Walter McDougall, *The Heavens and the Earth: A Political History of the Space Age* (John Hopkins University Press 1997) 274.

[14]Jeremy Black, *The Cold War: A Military History* (Bloomsbury 2015) 93–95.

[15]Association for Diplomatic Studies, 'Moments in U.S. Diplomatic History: Negotiating the Limited Test Ban Treaty (LTBT)' <https://adst.org/2016/12/negotiating-limited-test-ban-treaty-ltbt/> accessed 23 June 2018 | Office of the Historian, 'The Cuban Missile Crisis, October 1962' <https://history.state.gov/milestones/1961-1968/cuban-missile-crisis> accessed 23 June 2018.

[16]Iina Kohonen, 'The Space Race and Soviet Utopian Thinking' (2009) 57 The Sociological Review 114, 114 | Rip Bulkeley, *The Sputniks Crisis and Early United States Space Policy* (Macmillan 1991) 104–122.

[17]Peter Jankowitsch, 'The Background and History of Space Law' in Frans von der Dunk and Fabio Tronchetti (eds) *Handbook of Space Law* (Edward Elgar Publishing 2015) 2–3.

[18]Peter Jankowitsch, 'The Background and History of Space Law' in Frans von der Dunk and Fabio Tronchetti (eds) *Handbook of Space Law* (Edward Elgar Publishing 2015) 3.

spy satellite was launched within 2 years of Sputnik[19]), but outer space remained relatively free of an arms race and the achievements that were celebrated in the early years were predominantly civilian.

Following the launch of Sputnik 1 and Explorer 1, the USSR was once more ahead of the U.S. with the Luna 2 and Luna 3 missions, hitting the Moon and photographing the dark side of the Moon for the first time, respectively. Thereafter, the U.S. was nearly the first to achieve human spaceflight, but the Freedom 7 mission that would launch Alan Shepard into a sub-orbital trajectory was postponed and would only be successfully launched 1 month after Vostok 1.[20] Instead, the USSR's Vostok 1 mission achieved the next feat in the space race by successfully putting Yuri Gagarin into orbit around the Earth for a total duration of 108 min. Moreover, Valentina Tereshkova (Vostok 6) became the first woman in outer space and Alexey Leonov (Voshkod 2) conducted the first extravehicular activity (spacewalk).[21] Both the USSR and the U.S. achieved further 'firsts', such as the first Venus and Mars fly-by's, but the aforementioned are some of the most iconic 'firsts'. Ultimately, the most iconic feat during the space race was the Apollo 11 mission. Neil Armstrong and Buzz Aldrin became the first humans to land and walk on the Moon (with Michael Collins as the third astronaut of the Apollo 11 mission orbiting the Moon). This accomplishment was seen as the culmination of the space race.

Although many more 'firsts' would be achieved after the Moon landing, the thawing of the Cold War led to a new development in U.S.–USSR space exploration, namely cooperation. In 1975, they cooperated on the Apollo-Soyuz Test Project, or Soyuz-Apollo, docking an Apollo module and a Soyuz module.[22] The finest example of cooperation between the two space powers, however, is the International Space Station (ISS). Originally a cooperation between the U.S., Europe, Japan and Canada, Russia joined the ISS after the dissolution of the USSR.[23] Article 1 of the ISS Intergovernmental Agreement stipulates the objective of the agreement:

> to establish a long-term international cooperative framework among the Partners, on the basis of genuine partnership, for the detailed design, development, operation, and utilization of a permanently inhabited civil international Space Station for peaceful purposes, in accordance with international law. This civil international Space Station will enhance the scientific, technological, and commercial use of outer space.[24]

---

[19]Hannes Mayer, 'A Short Chronology of Spaceflight' in Christian Brünner and Alexander Soucek (eds) *Outer Space in Society, Politics and Law* (SpringerWienNewYork 2011) 22.

[20]Mike Wall, 'Space Race: Could the U.S. Have Beaten the Soviets into Space?' (Space.com, 8 April 2011) <https://www.space.com/11336-space-race-united-states-soviets-spaceflight-50years.html> accessed 6 November 2018.

[21]James Andrews and Asif Saddiqi (eds), *Into the Cosmos: Space Exploration and Soviet* Culture (University of Pittsburgh Press 2011) 195, 237.

[22]Edward Ezell and Linda Neuman Ezell, *The Partnership: A NASA History of the Apollo-Soyuz Test Project* (Dover Publications 2010).

[23]John Krige, *Fifty Years of European Cooperation in Space: Building on its Past, ESA shapes the Future* (Beauchesne 2014) 319–338.

[24]Agreement Among the Government of Canada, Governments of Member States of the European Space Agency, the Government of Japan, the Government of the Russian Federation, and the

The ISS thus exemplifies international cooperation between nearly all major space powers for peaceful purposes. However, it does not mean that all tension between the U.S. and Russia has dissipated. Apart from the obvious tension in general international affairs (Crimea, election hacking allegations, etc.), the U.S. and Russian positions in COPUOS often clash. Russia, for example, heavily protested the U.S. position on the exploitation of natural resources in outer space and the U.S. Commercial Space Launch Competitiveness Act of 2015.[25] Therefore, although the Cold War has ended, tension between the U.S. and Russia still remains, both in the CD and in COPUOS.

3.1.2 European Integration and Space Cooperation

The Second World War left Europe devastated and divided. Germany and Eastern Europe in particular had suffered enormous loses in population and destruction of infrastructure, manufacturing and economy.[26] France was faced with its own reconstruction, and the financial burden of the war had depleted the exchequer of the UK (as exemplified when the UK could not financially support Turkey during the Turkish Straits Crisis or the Greek Government in the Greek Civil War).[27] The power vacuum left by the defeat of Germany was filled by the U.S. and the USSR, both establishing their respective spheres of influence.[28] As a result, Europe was separated by an 'Iron Curtain', so called by Winston Churchill,[29] with the Warsaw Pact in the east and the North Atlantic Treaty Organization (NATO) in the west.[30] Only a few states in Europe, *i.e.* Austria, Switzerland, Sweden, Finland and Ireland, remained neutral. While the USSR began consolidating Eastern Europe into a communist bloc, the U.S., concerned about the communist ideology taking hold in Western Europe, started supporting the reconstruction of the Western European economies through the Marshall Plan.[31]

As the Marshall Plan helped the Western European economies recover, the Western European states started looking towards integrating facets of their

---

Government of the United States of America Concerning Cooperation on the Civil International Space Station (adopted 29 January 1998, entered into force 27 March 2001) TIAS 12927.

[25]U.S. Commercial Space Launch Competitiveness Act of 2015, H.R.2262—114th Congress (2015–2016), Public Law No. 114-90.

[26]Melvyn Leffler and David Painter (eds), *Origins of the Cold War: An International History* (2nd Edition Routledge 2005) 3.

[27]John Lamberton Harper, *The Cold War* (Oxford University Press 2011) 4 | Melvyn Leffler, *For the Soul of Mankind: The United States, The Soviet Union and the Cold War* (Hill and Wang 2007) 61 | Jeremy Black, *The Cold War: A Military History* (Bloomsbury 2015) 42–43.

[28]John Lamberton Harper, *The Cold War* (Oxford University Press 2011) 63–81.

[29]John Lamberton Harper, *The Cold War* (Oxford University Press 2011) 60.

[30]Jeremy Black, *The Cold War: A Military History* (Bloomsbury 2015) 45–47.

[31]Jeremy Black, *The Cold War: A Military History* (Bloomsbury 2015) 45–46.

economies to achieve peaceful cooperation.[32] The Schuman Plan (named after French Foreign Minister Robert Schuman) initially focused on applying the same policy for both the French and German coal and steel sectors but extended an open invitation to other states to participate in the project. This led to the creation of the European Coal and Steel Community (ECSC) in 1951, which included Germany, France, Italy, Belgium, the Netherlands and Luxembourg.[33] Another two communities to further European cooperation and integration were created within the decade, namely the European Economic Community (EEC), establishing the common market and harmonising economic policies between member states, and the European Atomic Energy Community (EAEC), dealing with common policies on nuclear energy.[34] These communities were brought under a single administration by the so-called Merger Treaty in 1967.[35]

To achieve even further integration of the European states, economically and monetarily, the European Union (EU) was created through the Maastricht Treaty.[36] However, due to continuous enlargement of the membership of the EU, further changes were needed to the institutions and the framework. This resulted in the adoption of the Lisbon Treaty in 2007, which created a new EU, with legal personality, to replace the communities (with the exception of the EAEC) and the EU created through the Maastricht Treaty.[37] Although the EU has recently been challenged by the European debt crisis and Brexit, among others, for the time being European integration remains the status quo and may increase.

Alongside the European integration that would eventually lead to the EU, European states began collaborating on space science and technology. Following the launch of Sputnik and the American response thereto, European states recognised that they were falling behind in the development of such technologies.[38] Europe responded by establishing the *Groupe d'études européen pour la collaboration dans la domaine des recherches spatiales* (GEERS), whose work led to the establishment of the European Space Research Organisation (ESRO), while the

---

[32] Koen Lenaerts and Piet van Nuffel, *European Union Law* (3rd edition Sweet and Maxwell 2011) 4.

[33] Athina Zervoyianni, George Argiros and George Agiomirgianakis, *European Integration* (Palgrave Macmillan 2006) 2–3 | Koen Lenaerts and Piet van Nuffel, *European Union Law* (3rd edition Sweet and Maxwell 2011) 8.

[34] Athina Zervoyianni, George Argiros and George Agiomirgianakis, *European Integration* (Palgrave Macmillan 2006) 4 | Koen Lenaerts and Piet van Nuffel, *European Union Law* (3rd edition Sweet and Maxwell 2011) 12–16.

[35] Athina Zervoyianni, George Argiros and George Agiomirgianakis, *European Integration* (Palgrave Macmillan 2006) 5.

[36] Athina Zervoyianni, George Argiros and George Agiomirgianakis, *European Integration* (Palgrave Macmillan 2006) 7–9 | Koen Lenaerts and Piet van Nuffel, *European Union Law* (3rd edition Sweet and Maxwell 2011) 39–42.

[37] Koen Lenaerts and Piet van Nuffel, *European Union Law* (3rd edition Sweet and Maxwell 2011) 65–67.

[38] John Krige, *Fifty Years of European Cooperation in Space: Building on its Past, ESA shapes the Future* (Beauchesne 2014) 15.

European Launcher Development Organisation (ELDO) was established in parallel negotiations.[39] As the names of the organisations already suggest, ESRO was responsible for space research and ELDO for launcher development. That these activities were separated and conducted by separate organisations was for both financial and political reasons: the costs of launcher development might have cut into the budget of space research, the costs would have been too high for smaller states and the launchers could have jeopardised the neutrality of Switzerland and Sweden.[40]

Early in the existence of ESRO and ELDO, however, it became clear that a more coordinated effort was necessary.[41] The need for a more coordinated effort led to the creation of the European Space Agency (ESA).[42] This idea of a more coordinated effort and stronger collaboration found its way into the ESA Convention.[43] The preamble, for example, reiterates an earlier resolution: 'that the aim would be to integrate the European national space programmes into a European space programme as far and as fast as reasonably possible'. This aim is also explicitly stipulated in Article II ESA Convention, along with the means to achieve the purpose of ESA:

> to provide for and to promote, for exclusively peaceful purposes, **cooperation (emphasis added)** among European states in space research and technology and their space applications, with a view of their being used for scientific purposes and for the operational space applications systems.

It is important to note that where the EU is an integration project, ESA is a cooperation project. This distinction brings a very different dynamic to the organisation and the position of Member States within it. Historically, ESA was created as an independent institution.[44] Interaction between ESA and the EU was thus limited. The Lisbon Treaty, however, made drastic changes by explicitly specifying EU

---

[39] John Krige, *Fifty Years of European Cooperation in Space: Building on its Past, ESA shapes the Future* (Beauchesne 2014) 19–21.

[40] John Krige, *Fifty Years of European Cooperation in Space: Building on its Past, ESA shapes the Future* (Beauchesne 2014) 20.

[41] John Krige, *Fifty Years of European Cooperation in Space: Building on its Past, ESA shapes the Future* (Beauchesne 2014) 169 | Thomas Hörber, 'Chaos or Consolidation? Post-war Space Policy in Europe' in Thomas Hörber and Paul Stephenson (eds) *European Space Policy: European Integration and the Final Frontier* (Routledge 2016) 27.

[42] Thomas Hörber, 'Chaos or Consolidation? Post-war Space Policy in Europe' in Thomas Hörber and Paul Stephenson (eds) *European Space Policy: European Integration and the Final Frontier* (Routledge 2016) 21–23.

[43] Convention for the Establishment of a European Space Agency (adopted 30 May 1975, entered into force 30 October 1980).

[44] Thomas Hörber, 'The European Space Agency and the European Union' in Thomas Hörber and Paul Stephenson (eds) *European Space Policy: European Integration and the Final Frontier* (Routledge 2016) 53–54.

competency on European space policy.[45] Article 189 of the Treaty on the Functioning of the European Union stipulates:

1. To promote scientific and technical progress, industrial competitiveness and the implementation of its policies, ***the Union shall draw up a European space policy (emphasis added)***. To this end, it may ***promote joint initiatives, support research and technological development and coordinate the efforts needed (emphasis added)*** for the exploration and exploitation of space.
2. To contribute to attaining the objectives referred to in paragraph 1, the European Parliament and the Council, acting in accordance with the ordinary legislative procedure, shall establish the necessary measures, which may take the form of a ***European space programme, excluding any harmonisation of the laws and regulations of the Member States (emphasis added)***.
3. The Union shall establish ***any appropriate relations with the European Space Agency (emphasis added)***.[46]

The emphases added to the article illustrate the extent of the competency of the EU pertaining to space policy. Although the EU is to draw up a European space policy through promotion, support and coordination, it is to do so without harmonisation of laws or regulations of member states. Furthermore, the article explicitly refers to the relationship with ESA.

Since the adoption of the Lisbon Treaty, EU space policy has been, and still is, under development, but the distribution of the work and responsibilities remains uncertain.[47] This leaves Europe with two vastly different organisations, ESA as an intergovernmental organisation and the EU as a supranational organisation involved in creating and steering European space policy. Although ESA member states have declared that ESA should not undergo changes to its intergovernmental nature or approach to space research,[48] it remains to be seen how the role of the EU in space develops. European cooperation in space science and technology naturally will have had an effect on the position of European states on some issues in COPUOS and the CD. As the European states have increased their cooperation and pursued common goals together, their views on issues naturally have become more aligned. The impact of ESA perhaps might have been minimal and limited to space matters that fall within the competency of ESA, but Article II(a) ESA Convention stipulates that ESA implements a long-term European space policy, and Article II(c) ESA Convention stipulates the coordinating function of ESA. The creation of an EU space policy has an even stronger potential to further align European stances. Although harmonisation of laws and regulations is excluded in the pursuit of creating a

---

[45]Koen Lenaerts and Piet van Nuffel, *European Union Law* (3rd edition Sweet and Maxwell 2011) 420–421.

[46]Consolidated version of the Treaty on the Functioning of the European Union (2016) OJ C202/1.

[47]Lucia Marta and Paul Stephenson, 'Role of the European Commission in Framing European Space Policy' in Thomas Hörber and Paul Stephenson (eds) *European Space Policy: European Integration and the Final Frontier* (Routledge 2016) 105.

[48]Lucia Marta and Paul Stephenson, 'Role of the European Commission in Framing European Space Policy' in Thomas Hörber and Paul Stephenson (eds) *European Space Policy: European Integration and the Final Frontier* (Routledge 2016) 105.

European space programme, an EU space policy and a European space programme will still coordinate the positions taken by EU member states on space matters. The extent of conformity among EU member states, however, remains to be seen.

3.1.3 The People's Republic of China's Position in the UN

The end of the Second World War saw the creation of the UN through the adoption and ratification of the UN Charter.[49] However, during the deliberation on the UN and UN Charter, China was not a unified state. Rather, both the nationalist and communist governments claimed authority over China.[50] However, the Nationalist government led by Chiang Kai-shek had jurisdiction over most of mainland China and Taiwan and was recognised as the representative for China by the major powers; the U.S., USSR and UK.[51] This led to Article 23 UN Charter stipulating that the Republic of China (ROC) was one of the five permanent members of the UN Security Council (UNSC). However, the Chinese civil war between the Nationalists and Communists continued, which culminated in the victory of the Communists and the establishment of the People's Republic of China (PRC) in 1949, while the ROC established itself in Taiwan but maintained its claim in mainland China. In the same year, the prime minister of the PRC, Chou En-lai, sent a telegram to the UN Secretary-General, demanding that the UN should transfer the Chinese seat on the UNSC to the PRC 'according to the principles and spirit of the Charter'.[52] This request did not result in any further action by the UNSC.[53] In 1950, Chou En-lai sent another telegram in which he protested the failure to expel the representative sent by the Chiang Kai-shek government.[54] Subsequently, the first proposal, which was rejected after deliberation, to reject and expel the Chinese Nationalist representatives in the UN was proposed by the USSR.[55]

---

[49] Charter of the United Nations (adopted 26 June 1945, entered into force 24 October 1945), see the preamble, which states: 'Accordingly, our respective Governments, through representatives assembled in the city of San Francisco, who have exhibited their full powers found to be in good and due form, have agreed to the present Charter of the United Nations and do hereby establish an international organization to be known as the United Nations'.

[50] Adriana Erthal Abdenur, 'Emerging Powers and the Creation of the UN: Three Ships of Thesus' (2016) 37 Third World Quarterly 1171, 1174.

[51] Adriana Erthal Abdenur, 'Emerging Powers and the Creation of the UN: Three Ships of Thesus' (2016) 37 Third World Quarterly 1171, 1178–1179.

[52] Evan Luard, 'China and the United Nations' (1971) 47 International Affairs 729, 729.

[53] Evan Luard, 'China and the United Nations' (1971) 47 International Affairs 729, 730.

[54] Evan Luard, 'China and the United Nations' (1971) 47 International Affairs 729, 729.

[55] Evan Luard, 'China and the United Nations' (1971) 47 International Affairs 729, 729.

In 1961, the UNGA through the adoption of Resolution 1668 (XVI) recognised China's representation in the UN as an important issue, which remained to be solved.[56] Despite the establishment of a committee to consider the issue,[57] the ROC maintained the Chinese seat and delivered speeches on behalf of China. For example, in 1959 during the 1080th meeting of the First Committee, the ROC stressed that it was imperative to use outer space for peaceful purposes, by making an analogy to the use of atomic energy.[58] Furthermore, the ROC stated that the primary objective of the UN in outer space was to ensure free and orderly use for peaceful purposes only and for the benefit of mankind.[59] Lastly, the ROC voted in favour of the establishment of the *ad hoc* Committee on the Peaceful Uses of Outer Space.[60]

It was not until 1971, through the adoption of Resolution 2758 (XXVI),[61] that the PRC was recognised as 'the only lawful representative of China to the United Nations and that the People's Republic of China is one of the five permanent members of the Security Council' and thus took up the Chinese seat. Resolution 2758 (XXVI) at the same time stipulated the expulsion of 'the representatives of Chiang Kai-shek from the place which they unlawfully occupy at the United Nations and in all organizations related to it'.[62]

The year before, the PRC had launched its first satellite, becoming only the fifth state to successfully demonstrate having independent access to outer space.[63] This event also tremendously increased incentives of the PRC to participate in the international dialogues and discussion of space matters. Up until that moment, the PRC had not played an active role in the discussion on space matters because of the legitimacy issues within the UN and also internal issues within the PRC government and its relative underdevelopment in the technological field and economy before 1971. The PRC thus did not make any statements in COPUOS (including on its mandate) until it joined the Committee in 1980.[64] In 1980, the PRC made a statement with the intention of gaining membership of COPUOS. However, COPUOS did not

---

[56]UNGA Res 1668 (XVI) (15 December 1961).

[57]Evan Luard, 'China and the United Nations' (1971) 47 International Affairs 729, 731.

[58]UNGA First Committee (14th Session) 'Summary Record of the 1080th Meeting' (11 December 1959) UN Doc A/C.1//SR.1080, par. 52.

[59]UNGA First Committee (14th Session) 'Summary Record of the 1080th Meeting' (11 December 1959) UN Doc A/C.1/SR.1080, par. 55.

[60]UNGA 'Record of the 792nd Plenary Meeting' UN GAOR 13th Session UN Doc A/PV.792 (13 December 1958), par. 175.

[61]UNGA Res 2758 (XXVI) (25 October 1971).

[62]UNGA Res 2758 (XXVI) (25 October 1971).

[63]Yun Zhao, 'National Space Legislation in Mainland China' (2007) 33 Journal of Space Law 427, 427 | UNGA Special Political Committee (35th Session) 'Summary Record of the 18th Meeting' (6 November 1980) UN Doc A/SPC/35/SR.18, par. 17.

[64]UNGA Res 35/16 (3 November 1980) UN Doc A/RES/35/16.

take a decision on the matter because it determined that such a decision fell outside its own mandate; instead, it was to be determined by the UNGA.[65] Subsequently, the PRC transmitted an official letter to the Special Political Committee to request admission to COPUOS.[66] The substantive part of the official letter reads as follows:

> I hereby formally communicate to the General Assembly that the Chinese Government wishes to apply for admission to membership in the Committee on the Peaceful Uses of Outer Space. Your kind co-operation and assistance in this respect would be greatly appreciated.[67]

This request was reiterated in the 18th meeting of the Special Political Committee.[68] China's request was taken up and resulted in a draft resolution entitled 'Admission of new members to the COPUOS', which addressed the need to expand the membership of COPUOS.[69] Although the U.S. requested a separate vote on section II of the Draft because of its negative attitude towards admitting Vietnam into COPUOS, the U.S. did reiterate its support for the PRC to become a member of COPUOS.[70] The request for a separate vote notwithstanding, draft resolution A/SPC/35/L.12 was adopted[71] and resulted in Resolution 35/16,[72] which officially recognised the PRC as a member of COPUOS.

The deliberations on the membership of the PRC illustrate that by 1980, the international community had recognised PRC's space capacity, for example through the statement of the U.S. that the PRC was an important space power.[73] Furthermore, the position of the PRC was quite exceptional because it was one of the few developing states that possessed independent access to outer space and had space capabilities. The following statement made by Argentina illustrates the importance of having a spacefaring developing state involved in the deliberations in COPUOS:

---

[65]UNGA 'Report of the Committee on the Peaceful Uses of Outer Space' UN GAOR 35th Session Supp 20 UN Doc A/35/20 (1980), 14.

[66]UNGA Special Political Committee (35th Session) 'Letter Dated 9 October 1980 from the Permanent Representative of China to the United Nations addressed to the President of the General Assembly' (27 October 1980) UN Doc A/SPC.35/4.

[67]UNGA Special Political Committee (35th Session) 'Letter Dated 9 October 1980 from the Permanent Representative of China to the United Nations addressed to the President of the General Assembly' (27 October 1980) UN Doc A/SPC.35/4.

[68]UNGA Special Political Committee (35th Session) 'Summary Record of the 18th Meeting' (6 November 1980) UN Doc A/SPC/35/SR.18, par. 17.

[69]UNGA Special Political Committee (35th Session) 'Summary Record of the 17th Meeting' (7 November 1980) UN Doc A/SPC/35/SR.17, par. 4.

[70]UNGA Special Political Committee (35th Session) 'Summary Record of the 19th Meeting' (10 November 1980) UN Doc A/SPC/35/SR.19, par. 35.

[71]UNGA Special Political Committee (35th Session) 'Summary Record of the 19th Meeting' (10 November 1980) UN Doc A/SPC/35/SR.19, par. 33–34.

[72]UNGA Res 35/16 (3 November 1980) UN Doc A/RES/35/16.

[73]UNGA Special Political Committee (35th Session) 'Summary Record of the 19th Meeting' (10 November 1980) UN Doc A/SPC/35/SR.19, par. 35.

> The developed countries are carrying out approximately 95 per cent of all research and development, whereas the developing countries, which make up 70 per cent of the world's population, have only 5 per cent of the world's capacity to conduct research and development activities and only the tiniest fraction of this small capacity is devoted to space activities and space applications.[74]

The enormous disparity between the developed and developing states emphasised the importance of the inclusion of the PRC in COPUOS.

Since the PRC started its reform policy in 1978, it has experienced strong economic development. Nowadays, the PRC's economy ranks as the second largest economy by nominal GDP and has the largest purchasing power according to the statistics of the International Monetary Fund (IMF).[75] This has also had its ramifications with respect to the PRC's space capabilities. In 2003, Shenzhou V was launched, which made the PRC the third country in the world to have independent human spaceflight capability.[76] Furthermore, the PRC has built its own navigation satellite system, namely the Beidou Navigation Satellite System.[77]

The foregoing demonstrates why the PRC historically has had a smaller role in the discussion of space matters internationally. This smaller role will be reflected in the smaller number of statements made, and thus included, in the research.

3.2 The Development of the Mandate of Committee on the Peaceful Uses of Outer Space (COPUOS)

Although Resolution 1348 (XIII) and Resolution 1472 (XIV) defined the mandate of COPUOS, the subsequent deliberations in COPUOS demonstrate the practical interpretation of this mandate. They also illustrate the further development and evolution of the interpretation of that mandate since its initial formulation. Accordingly, this section will examine, chronologically, the instruments discussed in COPUOS to delineate the evolution of the mandate of COPUOS and the discussions on the peaceful/military uses of outer space and the disarmament of outer space. By presenting this evolution, the conclusion of this chapter can delve into the interaction between COPUOS and the CD, the possible lack thereof and the need therefor.

---

[74] UNCOPUOS 'Verbatim Record of the Two Hundred and Twelfth Meeting' (14 July 1980) UN Doc A/AC.105/PV.212, 11.

[75] IMF, 'IMF Staff Completes 2018 Article IV Mission to China' (IMF, 29 May 2018) <https://www.imf.org/en/News/Articles/2018/05/29/pr18200-imf-staff-completes-2018-article-iv-mission-to-china> accessed 30 July 2018.

[76] Zheng Zhongyang, 'The Origins and Development of China's Manned Spaceflight Programme' (2007) 23 Space Policy 167, 170.

[77] Chunhao Han, Yuanxi Yang, Zhiwu Cai, 'BeiDou Navigation Satellite System and its Time Scales' (2011) 48 Metrologia 213.

3.2.1 The Development of the Mandate of COPUOS During the Negotiations of the UN Space Treaties

The development of international space law is usually divided into three distinct stages: the first stage consisting of non-legally binding resolutions, the second stage consisting of the development of international space law through the negotiation and adoption of international treaties and the third stage consisting of a return to non-legally binding 'soft law' resolutions and instruments.[78] This paragraph will discuss the development of the mandates during the second stage but will also include the negotiation of Resolution 1962 (XVIII). This resolution is included because it is the first international instrument on principles governing the activities of states in outer space, and the subsequent treaties are, for a large part, elaborations of the principles established in this resolution.

3.2.1.1 Resolution 1962 (XVIII): Declaration of Legal Principles Governing the Activities of States in the Exploration and Use of Outer Space

Immediately following the establishment of COPUOS, discussions on the peaceful and military uses of outer space resumed. This illustrates that certain military uses were still discussed in COPUOS despite the fact that the mandate referred to 'peaceful uses'. One of the primary goals of the Legal Subcommittee was to 'agree on the fundamental principles and to establish definite rules' applicable to the governance of outer space.[79] The USSR urgently embraced this task and suggested that the subcommittee should aim to prepare a declaration of the principles that should guide states in the exploration and use of outer space and a draft international agreement on assistance to and return of astronauts and space vehicles,[80] which should be conducted 'in a spirit of mutual understanding' and 'would yield results which would promote international co-operation in the exploration and use of outer space for peaceful purposes'.[81] Thus, there was once more a focus on the phrase 'peaceful purposes' without a clear definition. The U.S. clarified its interpretation of the mandate of COPUOS, stating:

> The United Nations Committee on the Peaceful Uses of Outer Space was directly concerned with international co-operation. That concern was separate from the subject matter of

---

[78]Frans von der Dunk, 'International Space Law' in Frans von der Dunk and Fabio Tronchetti (eds) *Handbook of Space Law* (Edward Elgar Publishing 2015) 37–43.
[79]UNCOPUOS (Legal Subcommittee), 'Summary Record of the First Meeting' (12 August 1962) UN Doc A/AC.105/C.2/SR.1, 4.
[80]UNCOPUOS (Legal Subcommittee), 'Summary Record of the First Meeting' (21 August 1962) UN Doc A/AC.105/C.2/SR.1, 6.
[81]UNCOPUOS (Legal Subcommittee), 'Summary Record of the First Meeting' (21 August 1962) UN Doc A/AC.105/C.2/SR.1, 7.

disarmament. Although ensuring that outer space was used solely for peaceful purpose was bound to involve agreements and measures in the context of disarmament, the Committee could nevertheless make significant collateral contributions to that end. [The U.S.] had always supported the principle that outer space should be used for peaceful purposes only – a goal which it believed could be reached only through measures of disarmament appropriately backed up by verification to ensure compliance (...).[82]

At first glance, the U.S. statement might seem to be inconsistent. First, the U.S. states that it considers that the discussion on disarmament falls outside the mandate of COPUOS, but then it recognises that ensuring the use of outer space exclusively for peaceful purposes requires 'agreements and measures in the context of disarmament'. The crux of the statement lies in the third sentence, which in essence states that the use of outer space for exclusively peaceful purposes can only be achieved through disarmament, but COPUOS can contribute to achieving that goal, not in the form of disarmament agreements but in the form of international cooperation. The statement, however, still leaves ambiguity as to whether other military uses of outer space fall within the mandate of COPUOS. It does, however, recognise that the issues are intertwined as it is acknowledged that disarmament is necessary to ensure that outer space will be used exclusively for peaceful purposes.

Other states were much clearer in their interpretation that military uses are to be discussed in COPUOS. Italy, for example, posed the question 'whether or not outer space should be closed to military use',[83] considering the Legal Subcommittee of COPUOS as the appropriate forum to discuss the issue. India then answered the Italian question in the affirmative, stating:

First, it would welcome a declaration by all Powers, and particularly by the two great Powers concerned, that outer space should be kept free from any military use. (...) The ultimate objective should be the conclusion of a convention aimed at the exclusively peaceful utilization of outer space for the benefit of man.[84]

The fact that India raised this issue in COPUOS indicates that it considered COPUOS the appropriate forum to discuss the military uses of outer space, or rather the prohibition of any military use of outer space.

Brazil took the complete opposite position. Not only should disarmament be discussed in the appropriate disarmament forum but so should 'The use of outer space for military purposes, the execution of nuclear tests in outer space, and the use of satellites for reconnaissance purposes'.[85] Brazil based this argument on 'the

---

[82] UNCOPUOS (Legal Subcommittee), 'Summary Record of the First Meeting' (21 August 1962) UN Doc A/AC.105/C.2/SR.1, 7.

[83] UNCOPUOS (Legal Subcommittee), 'Summary Record of the Second Meeting' (21 August 1962) UN Doc A/AC.105/C.2/SR.2, 2.

[84] UNCOPUOS (Legal Subcommittee), 'Summary Record of the Second Meeting' (21 August 1962) UN Doc A/AC.105/C.2/SR.2, 4.

[85] UNCOPUOS (Legal Subcommittee), 'Summary Record of the Fourth Meeting' (21 August 1962) UN Doc A/AC.105/C.2/SR.4, 2–3.

political realities of the time', further stating that it would be useless 'to try to prevent such uses outside a disarmament agreement implemented under effective international control'.[86] The interpretation that even non-arms military uses of outer space fell within the competence of the disarmament forums was supported by Australia. Australia gave the examples of the use of outer space for war propaganda and a general prohibition on any military use of outer space as issues that fell outside the mandate of COPUOS.[87] Moreover, Australia considered the issue of the military use of outer space as political rather than legal, which would make the Legal Subcommittee the wrong forum to discuss the issue regardless.[88]

The statements illustrate the ambiguous nature of the mandate given to COPUOS and the different interpretations of that mandate. Italy and India clearly interpreted the mandate to include a discussion of military uses of outer space, whereas Brazil and Australia regarded the disarmament framework as the appropriate forum to discuss such uses. The U.S. interpretation seems to favour the latter approach but is not clear enough to unequivocally put it in that bloc. Nevertheless, military uses of outer space were substantively discussed in COPUOS, for example, through the discussion, prompted by the USSR, of high-altitude nuclear tests.[89] This is furthermore evident from the USSR draft declaration of the basic principles governing the activities of states pertaining to the exploration and use of outer space.[90] The inclusion of a prohibition on military uses of outer space in the declaration was deemed as essential by the USSR, having already put forward such proposals in the Eighteen-Nation Committee,[91] despite the USSR stating:

> The United States representative had said realistically that a decision on that matter could be reached only as part of controlled disarmament. He agreed that prohibition of the use of outer space for military purposes did not come within the competence of the Sub-Committee, which should concentrate on other important matters.[92]

---

[86]UNCOPUOS (Legal Subcommittee), 'Summary Record of the Fourth Meeting' (21 August 1962) UN Doc A/AC.105/C.2/SR.4, 2–3.
[87]UNCOPUOS (Legal Subcommittee), 'Summary Record of the Fourth Meeting' (21 August 1962) UN Doc A/AC.105/C.2/SR.4, 4–5.
[88]UNCOPUOS (Legal Subcommittee), 'Summary Record of the Fourth Meeting' (21 August 1962) UN Doc A/AC.105/C.2/SR.4, 4–5.
[89]UNCOPUOS (Legal Subcommittee), 'Summary Record of the Fourth Meeting' (21 August 1962) UN Doc A/AC.105/C.2/SR.4, 8–14.
[90]UNCOPUOS, 'Union of Soviet Socialist Republics: Draft Declaration of the Basic Principles Governing the Activities of States Pertaining to the Exploration and Use of Outer Space' (10 September 1962) UN Doc A/AC.105/L.2.
[91]UNCOPUOS (Legal Subcommittee), 'Summary Record of the Seventh Meeting' (21 August 1962) UN Doc A/AC.105/C.2/SR.7, 4.
[92]UNCOPUOS (Legal Subcommittee), 'Summary Record of the Seventh Meeting' (21 August 1962) UN Doc A/AC.105/C.2/SR.7, 4.

Thus, although the USSR deemed a prohibition on the military uses of outer space essential, it did agree with the U.S. that such a prohibition should be part of controlled disarmament and that it did not fall within the mandate of the COPUOS Legal Subcommittee. The USSR position on the mandate of COPUOS is therefore hard to gauge, particularly because the earliest USSR drafts included two substantive provisions on the military uses of outer space—a provision that 'the use of outer space for propagating war, national or racial hatred or enmity between nations shall be prohibited'[93] and a provision prohibiting the use of satellites for collecting intelligence information (spy satellites).[94]

These paragraphs met strong resistance from the U.S. With respect to paragraph 5, the U.S. referred to a similar principle that had already been agreed upon by the USSR in the Disarmament Committee but was eventually repudiated by the USSR.[95] Therefore, the U.S. felt that it was disingenuous to include the principle in the draft declaration. The opposition towards paragraph 8 stemmed from the idea that 'international law imposed no prohibition on the observation of the earth from outer space, which was peaceful and did not interfere with other activities on earth or in space', adding that 'indeed any other observation which the USSR might be conducting in outer space, were peaceful and that Major Titov's military status and the intent of his observations were irrelevant'.[96] The U.S. thus did not view the collection of information from outer space as inherently military. In contrast, the USSR made a distinction between the military use of Earth observation (EO) and civil use, asserting that military use, or espionage, was undesirable, while the gathering of scientific data should be allowed.[97]

Following the USSR draft declaration, two factions can be discerned. The first agreed with the U.S. that in addition to the disarmament of outer space, military uses of outer space should be discussed in the disarmament framework. This faction was exemplified by statements made by France and Australia. France, for example, stated:

---

[93] UNCOPUOS 'Union of Soviet Socialist Republics: Draft Declaration of the Basic Principles Governing the Activities of States Pertaining to the Exploration and Use of Outer Space' (10 September 1962) UN Doc A/AC.105/L.2, par. 5.

[94] UNCOPUOS 'Union of Soviet Socialist Republics: Draft Declaration of the Basic Principles Governing the Activities of States Pertaining to the Exploration and Use of Outer Space' (10 September 1962) UN Doc A/AC.105/L.2, par. 8.

[95] UNCOPUOS (Legal Subcommittee) 'Summary Record of the Seventh Meeting' (21 August 1962) UN Doc A/AC.105/C.2/SR.7, 8.

[96] UNCOPUOS (Legal Subcommittee), 'Summary Record of the Seventh Meeting' (21 August 1962) UN Doc A/AC.105/C.2/SR.7, 9 I Major Titov was the second human to orbit Earth and made photos of the Earth from outer space. The 'observations' referred to by the U.S. are these photos. The U.S. compared such observations, already conducted by the USSR to observations made by satellites.

[97] UNCOPUOS (Legal Subcommittee), 'Summary Record of the Seventh Meeting' (21 August 1962) UN Doc A/AC.105/C.2/SR.7, 12.

The reference to war propaganda in principle 5 and the use of satellites for the collection of intelligence in principle 8 was out of place in such a declaration. Disarmament problems were outside the competence of the Sub-Committee, whose task – laying the foundations of the law of outer space – was already vast enough and should not be complicated by the introduction of extraneous matter.[98]

This position, also supported by Canada and the United Kingdom,[99] thus interpreted the use of outer space for war propaganda or the collection of intelligence information as a disarmament problem. Therefore, it was outside the mandate of COPUOS to deliberate on these issues. Australia voiced support for this position when considering the mandate of the Legal Subcommittee:

> First, however, it was necessary to determine its proper tasks; [Australia] could not agree that the Sub-Committee's mandate was comprehensively legislative in the sense that it could deal with any subject whatever and recommend what the law should be. Other organs were responsible for considering some aspects of the regime of outer space. *The Eighteen-Nation Disarmament Committee, for example, was dealing with matters of disarmament which were outside the Sub-Committee's term of reference (emphasis added)*. It was proper that where relevant the Sub-Committee's decisions should wait on those of the Eighteen-Nation Disarmament Committee.
>
> (...)
>
> On the other hand, *some of the matters dealt with in principles 5, 6, 7 and 8 of the Soviet draft declaration seemed to be plainly within the competence of other bodies, such as the Disarmament Committee. (emphasis added)* All four clauses included elements which required decisions wholly political in character and which could not usefully be considered by the Sub-Committee until the necessary political decisions had been taken by the appropriate organ of the United Nations.[100]

Australia thus strongly stated that the mandate of COPUOS and its Legal Subcommittee did not extend to the discussion of disarmament issues. Furthermore, by explicitly mentioning principles 5 and 8 in the context of the Disarmament Committee, Australia implied that such uses of outer space fell within the discussion of the disarmament of outer space.

The second faction, following the USSR interpretation, held that a distinction should be made between true disarmament issues, which should be discussed in the disarmament framework, and military uses of outer space that were not disarmament issues, which should be discussed in COPUOS. The USSR based this interpretation on the fact that Resolution 1721 (XVI) had recommended the application of general principles of international law applicable to the exploration and use of outer space. These principles, however, would need to be supplemented by further principles to

---

[98]UNCOPUOS (Legal Subcommittee), 'Summary Record of the Ninth Meeting' (21 August 1962) UN Doc A/AC.105/C.2/SR.9, 3.

[99]UNCOPUOS (Legal Subcommittee), 'Summary Record of the Sixth Meeting' (21 August 1962) UN Doc A/AC.105/C.2/SR.9, 7 | UNCOPUOS (Legal Subcommittee), 'Summary Record of the Tenth Meeting' (21 August 1962) UN Doc A/AC.105/C.2/SR.10, 3–4.

[100]UNCOPUOS (Legal Subcommittee), 'Summary Record of the Eleventh Meeting' (21 August 1962) UN Doc A/AC.105/C.2/SR.11, 6–7.

be discussed in the Legal Subcommittee.[101] One of the general principles in Resolution 1721 (XVI) was the applicability of general international law, including the UN Charter, to outer space. The USSR thus saw it as the task of COPUOS to elaborate on this principle, including the prohibition of military uses of outer space as an extension of Article 2(4) UN Charter. This position is exemplified by a statement made by Romania:

> Doubts had been raised as to the Sub-Committee's competence; it had been suggested that the prohibition of aggression, for example, was more properly a topic to be discussed in the Eighteen-Nation Disarmament Committee. Yet it was not impossible that that Committee would fail to reach any conclusion on the matter. In any event, ***the ban on aggression did not come within the context of disarmament; rather it was the subject of a rule of general international law which was, moreover, enshrined in the United Nations Charter. (emphasis added)*** The Sub-Committee was not expected to frame new rules, but to apply existing rules to outer space. He could not accept the argument that political and legal aspects should be separated, and that before the Sub-Committee could act some other body should take a political decision. The Sub-Committee was composed of acknowledged specialists sitting, not as experts but as representative of sovereign States, members of the United Nations, and therefore competent to establish principles of international law.[102]

The USSR position thus viewed the prohibition on the military use of outer space not as a disarmament issue but rather as a matter of general international law. This position was supported not just by the USSR and its Eastern European allies, but also by Italy. Italy called for a complete ban on the use of outer space for military purposes because it 'was a special case needing careful consideration'.[103] The context surrounding this statement and the earlier statements made by Italy indicate that Italy considered that the discussion on military uses should be included in the deliberations of COPUOS. The general discussion on the need to adopt a declaration on principles applicable to outer space activities, or rather the lack of agreement on these issues specifically, eventually led to the conclusion of the 1962 session of the Legal Subcommittee.[104]

Nevertheless, the discussion on whether or not to deny the use of outer space for observation and photography, and the inclusion of a provision prohibiting war propaganda, continued in the plenary meeting of COPUOS.[105] The USSR held on to its view that observation and photography had military dimensions and as such

---

[101] UNCOPUOS (Legal Subcommittee), 'Summary Record of the Fourteenth Meeting' (22 August 1962) UN Doc A/AC.105/C.2/SR.14, 2–3.

[102] UNCOPUOS (Legal Subcommittee), 'Summary Record of the Thirteenth Meeting' (22 August 1962) UN Doc A/AC.105/C.2/SR.13, 2.

[103] UNCOPUOS (Legal Subcommittee), 'Summary Record of the Sixth Meeting' (21 August 1962) UN Doc A/AC.105/C.2/SR.7, 7 & 12.

[104] UNCOPUOS (Legal Subcommittee), 'Summary Record of the Fifteenth Meeting' (22 August 1962) UN Doc A/AC.105/C.2/SR.15.

[105] UNCOPUOS 'Verbatim Records of the Eleventh Meeting' (21 February 1963) UN Doc A/AC.105/PV.11, 6–7.

should be prohibited and that war propaganda should be banned.[106] At the same time, the U.S. remained firm in its position. The conflict between the U.S. and USSR positions, as well as the uncertainty about the USSR position, is exemplified by the following Canadian statement:

> The United States has taken the view, which is close to the view of [Canada], that problems relating to the disarmament of or the exclusion of arms from outer space should be dealt with in the eighteen-nation Disarmament Committee. The Soviet Union, however, appears to believe that certain measures of a disarmament character can be secured through our Committee. I say only 'certain measures of disarmament character' because, when the issue was raised directly in the Legal Sub-Committee in Geneva, the Soviet delegation maintained that substantial disarmament measures in outer space were the responsibility of the Eighteen-Nation Disarmament Committee.[107]

Therefore, Canada and the U.S. considered issues such as observation satellites, high-altitude nuclear tests and war propaganda as having a disarmament character and thus should be discussed in the disarmament framework.[108] This position was also embraced by France.[109] The USSR responded by stating that the USSR draft declaration did not impede the mandate of the Disarmament Committee; rather, the contested issues in the draft declaration are included to ensure that peaceful international cooperation can occur.[110] It was the USSR position that such peaceful international cooperation cannot occur if outer space was misused by carrying out nuclear tests, disseminating war propaganda and using satellites for espionage. In essence, the USSR thus stated that disarmament issues belonged in the disarmament framework but that certain military uses of outer space should be discussed in COPUOS because they related to the peaceful use of outer space or were a prerequisite for peaceful international cooperation. A further response from the USSR once more reiterated that the USSR regarded the launch of military satellites (or spy satellites) as inconsistent with the goal of having peaceful cooperation in the use of outer space.[111]

Eventually, COPUOS reported to the General Assembly that with regard to the work of the Legal Subcommittee, no agreement had been reached in the Legal

---

[106]UNCOPUOS 'Verbatim Records of the Eleventh Meeting' (21 February 1963) UN Doc A/AC.105/PV.11, 38–41.
[107]UNCOPUOS 'Verbatim Records of the Twelfth Meeting' (21 February 1963) UN Doc A/AC.105/PV.12, 15.
[108]UNCOPUOS 'Verbatim Records of the Twelfth Meeting' (21 February 1963) UN Doc A/AC.105/PV.12, 31.
[109]UNCOPUOS 'Verbatim Records of the Thirteenth Meeting' (21 February 1963) UN Doc A/AC.105/PV.13, 17.
[110]UNCOPUOS 'Verbatim Records of the Twelfth Meeting' (21 February 1963) UN Doc A/AC.105/PV.12, 17–18.
[111]UNCOPUOS 'Verbatim Records of the Twelfth Meeting' (21 February 1963) UN Doc A/AC.105/PV.12, 32.

Subcommittee or the Plenary.[112] As a response, the UNGA adopted Resolution 1802 (XVII), which spurred COPUOS on to continue its work on the elaboration of basic legal principles governing the activities of states in the exploration of outer space, on liability for space vehicle accidents and on assistance to and return of astronauts and space vehicles, as well as on other legal problems.[113]

Between the conclusion of the 1962 session of the Legal Subcommittee and the start of the 1963 session, strides were made, which were reflected in U.S. opening statement that it was ready to join in a further formulation of general principles of space law.[114] Although the USSR had submitted a new draft resolution on such general principles, this draft resolution still contained a prohibition on the use of outer space for war propaganda and a principle that the use of artificial satellites for the collection of intelligence information in the territory of a foreign state was incompatible with the objectives of mankind in its conquest of outer space.[115] These issues were kept in the USSR draft because it considered the use of satellites for telecommunication an effective means to propagating war and thus wanted it banned in accordance with earlier UNGA resolutions.[116]

Furthermore, the banning of espionage through satellites was included in the draft to convey the USSR understanding of norms of international law, maritime law and air law that already banned espionage activities. According to the USSR, this prohibition should be extended to outer space because the altitude from which intelligence observations might be made was immaterial.[117]

The U.S. responded by submitting its own draft resolution for a declaration on general legal principles.[118] In contrast to the USSR draft, the U.S. draft did not contain any reference to the banning of military uses of outer space, specific or general. Likewise, and in accordance with the U.S. perspective, France and Australia considered the issues of propaganda and information activities outside the mandate of the Legal Subcommittee.[119] The U.S. viewpoint found further support from the

---

[112] UNGA 'Report of the Committee on the Peaceful Uses of Outer Space' UN GOAR 17th Session UN Doc A/5181 (27 September 1962), 5.

[113] UNGA Res 1802 (XVII) (14 December 1962).

[114] UNCOPUOS (Legal Subcommittee), 'Summary Record of the Sixteenth Meeting' (27 June 1963) UN Doc A/AC.105/C.2/SR.16, 4.

[115] UNCOPUOS (Legal Subcommittee), 'Summary Record of the Seventeenth Meeting' (27 June 1963) UN Doc A/AC.105/C.2/SR.17, 6–7.

[116] UNCOPUOS (Legal Subcommittee), 'Summary Record of the Twenty-Second Meeting' (26 April 1963) UN Doc A/AC.105/C.2/SR.22, 4.

[117] UNCOPUOS (Legal Subcommittee), 'Summary Record of the Twenty-Second Meeting' (26 April 1963) UN Doc A/AC.105/C.2/SR.22, 4.

[118] UNGA First Committee (17th Session) 'Letter Dated 8 December 1962 from the representative of the United States of America to the Chairman of the First Committee' (8 December 1962) UN Doc A/C.1/881 | UNCOPUOS (Legal Subcommittee), 'Summary Record of the Twentieth Meeting' (27 June 1963) UN Doc A/AC.105/C.2/SR.20, 10.

[119] UNCOPUOS (Legal Subcommittee), 'Summary Record of the Twenty-Second Meeting' (26 April 1963) UN Doc A/AC.105/C.2/SR.22, 14 | UNCOPUOS (Legal Subcommittee), 'Summary Record of the Twenty-Third Meeting' (29 April 1963) UN Doc A/AC.105/C.2/SR.23, 6–7.

UK, which agreed that the consideration of peaceful uses of outer space and the implications of such a concept should be deliberated upon in the Eighteen-Nation Committee.[120] Just as the previous year, the Legal Subcommittee was divided, a divide that was recognised by Australia:

> Paragraph 1 of the United Arab Republic draft code (A/AC.105/L.5), in which it was provided that the activities of Member States in outer space should be confined to peaceful uses, had received some support in the Sub-Committee. On the other hand, it had been commented on adversely by the delegations which held that the question of military activities in outer space was not within the Sub-Committee's terms of reference and that general and complete disarmament was the only way to deal with the question.[121]

However, not all other members of COPUOS exactly adhered to the U.S. or the USSR interpretation. Italy, for example, leaned toward the USSR interpretation by maintaining the position that it had already taken in 1962, namely that all activities of an aggressive nature in outer space should be banned.[122] However, the term 'aggressive' is distinct from 'military' and only includes uses of outer space that are aggressive towards another state, such as the threat or the use of force,[123] while 'military' generally encompasses more uses. Therefore, Italy seems to have put forward a weakened position on this issue compared to its position in 1962.

This position was also supported by the United Arab Republic (UAR) draft resolution and statements made by India, which called for a provision or principle limiting the use of outer space to peaceful purposes.[124] India recognised that such a prohibition was connected to the question of disarmament but also considered it a necessary step in the development of international space law and a stepping stone to achieving general and complete disarmament.[125] To support its interpretation, India referred to the stance expressed by the USSR and the U.S. in earlier deliberations on the use of outer space exclusively for peaceful purposes, concluding that the banning of military uses of outer space and the issue of disarmament were distinct and could be discussed in parallel.[126] A similar view on the discussion of the peaceful uses of outer space and disarmament was taken by Japan. On the one hand, Japan recognised that the prohibition of military uses of outer space could only be achieved in the

---

[120]UNCOPUOS (Legal Subcommittee), 'Summary Record of the Twenty-Fourth Meeting' (1 May 1963) UN Doc A/AC.105/C.2/SR.24, 13.

[121]UNCOPUOS (Legal Subcommittee), 'Summary Record of the Twenty-Third Meeting' (29 April 1963) UN Doc A/AC.105/C.2/SR.23, 6.

[122]UNCOPUOS (Legal Subcommittee), 'Summary Record of the Twentieth Meeting' (27 June 1963) UN Doc A/AC.105/C.2/SR.20, 4.

[123]UNCOPUOS (Legal Subcommittee), 'Summary Record of the Twentieth Meeting' (27 June 1963) UN Doc A/AC.105/C.2/SR.20, 4.

[124]UNCOPUOS (Legal Subcommittee), 'Summary Record of the Twenty-Second Meeting' (26 April 1963) UN Doc A/AC.105/C.2/SR.22, 7.

[125]UNCOPUOS (Legal Subcommittee), 'Summary Record of the Twenty-Second Meeting' (26 April 1963) UN Doc A/AC.105/C.2/SR.22, 8.

[126]UNCOPUOS (Legal Subcommittee), 'Summary Record of the Twenty-Second Meeting' (26 April 1963) UN Doc A/AC.105/C.2/SR.22, 8.

framework of controlled disarmament and that such an agreement should be deliberated upon in the Eighteen-Nation Committee. On the other, Japan stated that COPUOS could guide such deliberations by giving guidance on the matter of the peaceful uses of outer space.[127] Finally, this stance, and the interpretation that COPUOS was mandated to discuss such issues, was also supported by Lebanon, which stated:

> Activities of Member States in outer space should be confined solely to the peaceful uses. In his opinion, that question was within the terms of reference of the Committee on the Peaceful Uses of Outer Space and need not be handed over to the disarmament negotiators.[128]

The end result of the discussion in the Legal Subcommittee, however, was that no real agreement was achieved and that further deliberations were needed to reach agreement on the issues.[129]

Once more, the discussion moved to the Plenary of COPUOS. However, an important shift occurred between the adoption of the report of the Legal Subcommittee and the start of deliberations in the Plenary because the U.S., the USSR and the UK adopted the Partial Test-Ban Treaty (PTBT). The adoption of this instrument, which was also discussed in COPUOS, allowed for a different atmosphere in the deliberations of COPUOS.[130] Notwithstanding, the USSR initially held its position on 'the impermissibility of the use of satellites for collecting intelligence information, for war propaganda, and for propaganda connected with national and racial hatred and enmity among peoples'.[131] Unsurprisingly, the U.S. did not change its position either and maintained that such issues should be discussed within the appropriate disarmament framework. Therefore, the deliberations in the Plenary did not result in an agreed draft.[132]

Although neither the Legal Subcommittee nor the Plenary came to a result on a declaration, continuing contacts and exchanges of views outside COPUOS eventually led to a draft declaration of legal principles governing the activities of states in the exploration and use of outer space.[133] This draft declaration omitted any reference to observation through satellites but did refer to war propaganda in a more

---

[127] UNCOPUOS (Legal Subcommittee), 'Summary Record of the Twenty-Second Meeting' (26 April 1963) UN Doc A/AC.105/C.2/SR.22, 11.

[128] UNCOPUOS (Legal Subcommittee), 'Summary Record of the Twenty-First Meeting' (25 April 1963) UN Doc A/AC.105/C.2/SR.21, 9.

[129] UNCOPUOS (Legal Subcommittee), 'Report of the Legal Sub-Committee on the Work of its Second Session (16 April–3 May 1963) to the Committee on the Peaceful Uses of Outer Space' (6 May 1963) UN Doc A/AC.105/12, 4.

[130] UNCOPUOS 'Verbatim Records of the Twelfth Meeting' (10 October 1963) UN Doc A/AC.105/PV.20, 17.

[131] UNCOPUOS 'Verbatim Records of the Twelfth Meeting' (10 October 1963) UN Doc A/AC.105/PV.20, 29.

[132] UNCOPUOS 'Report of the Committee on the Peaceful Uses of Outer Space' (24 September 1963) UN Doc A/5549 and ADD.1.

[133] UNCOPUOS 'Verbatim Records of the Twenty-Fourth Meeting' (22 November 1963) UN Doc A/AC.105/PV.24, 3.

general way in its preamble by making a reference to Resolution 110 (II), which condemned propaganda designed or likely to provoke or encourage any threat to the peace, breach of the peace or act of aggression and considered that the aforementioned resolution is applicable to outer space.[134]

The USSR addressed the absence of a provision on satellite observation, stating that there were still unsettled aspects for the law of outer space and that the USSR by no means considered the draft declaration of legal principles as exhaustive.[135] The USSR was even more explicit with respect to the mandate of COPUOS, stating:

> We cannot fail to observe that one group of the comments was directed towards prejudging the solution of certain problems falling within the ambit of outer space which can and should be solved within the framework of the problem of general and complete disarmament. As we have pointed on numerous occasions, the Soviet Union is prepared positively to solve these matters as well, but it cannot permit of their being divorced from the solution of other matters connected with them and related to disarmament.[136]

Following this short discussion in COPUOS, the discussion of the draft declaration proceeded in the First Committee. The inclusion of the preambular provision that stipulated that Resolution 110 (II)[137] on war propaganda was equally applicable to outer space was specifically mentioned by the USSR in the discussion of the draft declaration as an important concession.[138] In addition, the USSR stated:

> **The draft declaration did not, and indeed could not, touch the use of outer space for military purposes.** *(emphasis added)* The Soviet Union had repeatedly declared that it was prepared to destroy all types of armaments as part of a programme of general and complete disarmament under strict international control, which would ipso facto solve the problem of prohibiting the use of outer space for military purposes. The USSR could not agree to the separation of that problem from other disarmament measures directly related to it, such as the elimination of military bases in foreign territories.[139]

This statement indicated a reversal of the USSR interpretation of the mandate of COPUOS. Instead, it favoured the U.S. interpretation to keep the discussion of military uses of outer space outside of COPUOS and contained within the appropriate disarmament framework. However, it was also in contrast with the jubilation about the concession on the inclusion of war propaganda in the draft declaration. The USSR statements thus obfuscated the interpretation of the USSR on the mandate of COPUOS. Together with the earlier USSR statements, the USSR seems to have kept its distinction between non-arms military uses, such as war propaganda and satellite

---

[134]UNCOPUOS 'Additional Report' (27 November 1963) UN Doc A/5549/ADD.1.
[135]UNCOPUOS 'Verbatim Records of the Twenty-Fourth Meeting' (22 November 1963) UN Doc A/AC.105/PV.24, 51–55.
[136]UNCOPUOS 'Verbatim Records of the Twenty-Fourth Meeting' (22 November 1963) UN Doc A/AC.105/PV.24, 52.
[137]UNGA Res 110 (II) (3 November 1947) UN Doc A/RES/2/110.
[138]UNGA First Committee (18th Session) 'Summary Record of the 1342nd Meeting' (2 December 1963) UN Doc A/C.1/SR.1342, 161.
[139]UNGA First Committee (18th Session) 'Summary Record of the 1342nd Meeting' (2 December 1963) UN Doc A/C.1/SR.1342, 161.

observation, and arms military uses of outer space. The former then should be discussed in COPUOS, while the latter should be discussed in the disarmament framework.

In contrast, Australia supported a stronger distinction between the mandate of COPUOS and the disarmament forum or rather a stronger distinction about which activities fall within the scope of the disarmament framework, namely non-arms military uses such as espionage. Australia argued that although Resolution 1884 (XVIII) and the PTBT affected the work of COPUOS, it was not for COPUOS to deal with such issues.[140] Instead, the Eighteen-Nation Committee should deal with those matters, but 'it was only realistic to recognize that the spheres of interest of the two Committees did touch upon each other, even if they did not actually overlap'.[141]

Nevertheless, the statements of other states still demonstrate that neither the U.S. nor the USSR interpretation had a majority, nor were they the only interpretations. Austria, for example, stated that the draft declaration should have included a provision acknowledging or reiterating the resolution on prohibiting the placing of weapons of mass destruction in outer space.[142] This statement indicates that under Austria's interpretation, COPUOS would be mandated to discuss military uses of outer space. The UAR went further, referring to earlier statements of Japan, Lebanon, India and Brazil, as well as its own draft code, stating that outer space should be used solely for peaceful purposes similar to the Antarctic Treaty, which prohibited any measures of a military nature.[143] Therefore, the UAR also recognised COPUOS as the appropriate forum to discuss the military uses of outer space. India echoed this statement and referred both to confining the use of space to peaceful uses and to the absence of a reference to Resolution 1884 (XVIII), which called upon states to refrain from placing weapons of mass destruction in outer space.[144] With respect to the mandate, India made the following statement:

> Although the question of the peaceful uses of outer space *undeniably connected with that of disarmament and it was sometimes difficult to distinguish peaceful from military uses (emphasis added)*, the enunciation of [the principle that activities of states in outer space should be confined to peaceful uses would constitute a significant step in the development of the rule of law in outer space.
>
> (...)
>
> It had been argued at the seventeenth session by both the United States and the Soviet Union, first, that [COPUOS] was not competent to deal with the question of reserving outer space for peaceful uses, which was closely linked with the question of disarmament and therefore a

---

[140] UNGA First Committee (18th Session), 'Summary Record of the 1345th Meeting' (5 December 1963) UN Doc A/C.1/SR.1345, 184.

[141] UNGA First Committee (18th Session), 'Summary Record of the 1345th Meeting' (5 December 1963) UN Doc A/C.1/SR.1345, 184.

[142] UNGA First Committee (18th Session) 'Summary Record of the 1342nd Meeting' (2 December 1963) UN Doc A/C.1/SR.1342, 161.

[143] UNGA First Committee (18th Session) 'Summary Record of the 1342nd Meeting' (2 December 1963) UN Doc A/C.1/SR.1342, 162.

[144] UNGA First Committee (18th Session), 'Summary Record of the 1343rd Meeting' (3 December 1963) UN Doc A/C.1/SR.1343, 168.

Table 3.1 Interpretation of the mandate of COPUOS in 1963

	US	USSR	India
Armed military uses	Disarmament framework	Disarmament framework	Disarmament framework (but limiting the use for exclusively peaceful purposes can be done in COPUOS)
Non-arms military uses	Disarmament framework	COPUOS	COPUOS
Peaceful uses	COPUOS	COPUOS	COPUOS

matter for exclusive consideration by the Conference of the Eighteen-Nation Committee on Disarmament; and secondly, that the adoption of a legal principle relating to the military use of outer space would be contrary to the accepted policy followed in disarmament negotiations, inasmuch as there would be no provision for verification. [India] could not agree that [COPUOS] was not competent in the matter, since the Committee had come into being because of the space powers' concern to avoid any misuse of outer space.[145]

This statement summarises the positions of the U.S. and the USSR on the mandate of COPUOS but also demonstrates that other states had wholly different positions on this mandate. India argued that because states had already agreed that the exploration of outer space should be carried out through cooperation, it should be possible for states to agree to limiting the use of outer space exclusively to peaceful purposes and to agree to that outside the framework of general and complete disarmament, pointing out that Resolution 1884 (XVIII) on placing weapons of mass destruction in outer space had also been adopted without a verification mechanism.[146] Therefore, after the negotiation of Resolution 1962 (XVIII), which in the end did not contain any reference to military uses of outer space,[147] the various interpretations can be summarised as reproduced in Table 3.1 (the use of the U.S., the USSR and India to refer to the interpretations is a simplification to make a clear distinction between the different interpretations; of course, there were other states that also adhered to the various interpretations or variations of those interpretations).

3.2.1.2 The Treaty on Principles Governing the Activities of States in the Exploration and Use of Outer Space, Including the Moon and Other Celestial Bodies (OST)

Discussions in COPUOS and its Legal Subcommittee continued after the adoption of Resolution 1962 (XVIII) with the deliberations on a legally binding international instrument stipulating the principles governing activities conducted in outer space.

---

[145]UNGA First Committee (18th Session), 'Summary Record of the 1343rd Meeting' (3 December 1963) UN Doc A/C.1/SR.1343, 168.

[146]UNGA First Committee (18th Session), 'Summary Record of the 1343rd Meeting' (3 December 1963) UN Doc A/C.1/SR.1343, 168.

[147]UNGA Res 1962 (XVIII) (13 December 1963).

This discussion stemmed from Resolution 1963 (XVIII), which recommended 'that consideration should be given to incorporating in international agreement form, in the future as appropriate, legal principles governing the activities of States in the exploration and use of outer space'.[148]

In its opening statement at the Legal Subcommittee, the USSR stated that the Legal Subcommittee was the only UN body that dealt with the legal problems of outer space; therefore, it should consider the whole field of the law of outer space.[149] The USSR did not specify whether this should include military uses and disarmament or be limited to the field of the law of the *peaceful* uses of outer space instead. The focus on the fact that the COPUOS Legal Subcommittee is the *only* UN body dealing with legal problems of outer space and that it should consider the *whole* field of outer space indicates that it is the former rather than the latter.

The same states that wanted to include the discussion on military uses of outer space in the deliberations on Resolution 1962 (XVIII) brought up the subject again in the further discussion on an international agreement dealing with general principles. Japan stated that the general principles should be expanded as soon as possible to include the principle that outer space should be used for exclusively peaceful purposes.[150] Likewise, the UAR stated that the use of outer space for military purposes must be prohibited, adding: 'It was clearly the Sub-Committee's duty to press for the adoption of such a principle by the General Assembly'.[151] This position was further reiterated by Lebanon, which touched upon the necessity of a prohibition of military activities in outer space as that was the only way, in their eyes, that the exploration and use of outer space could be carried on for the benefit of mankind.[152]

When it came to drafting the report of the Legal Subcommittee, these same states raised their concern when the revised draft made no mention of their statements with respect to the exclusive use of outer space for peaceful purposes.[153] This point was reiterated by India in the closing of the first part of the session of the Legal

---

[148] UNGA Res 1963 (XVIII) (13 December 1963).

[149] UNCOPUOS (Legal Subcommittee), 'Summary Records of the 29th to 37th Meetings Held at the Palais des Nations, Geneva, from 9 to 26 March 1964' (24 August 1964) UN Doc A/AC.105/C.2/SR.29–37, 2.

[150] UNCOPUOS (Legal Subcommittee), 'Summary Records of the 29th to 37th Meetings Held at the Palais des Nations, Geneva, from 9 to 26 March 1964' (24 August 1964) UN Doc A/AC.105/C.2/SR.29–37, 22.

[151] UNCOPUOS (Legal Subcommittee), 'Summary Records of the 29th to 37th Meetings Held at the Palais des Nations, Geneva, from 9 to 26 March 1964' (24 August 1964) UN Doc A/AC.105/C.2/SR.29–37, 45.

[152] UNCOPUOS (Legal Subcommittee), 'Summary Records of the 29th to 37th Meetings Held at the Palais des Nations, Geneva, from 9 to 26 March 1964' (24 August 1964) UN Doc A/AC.105/C.2/SR.29–37, 48.

[153] UNCOPUOS (Legal Subcommittee), 'Summary Records of the 29th to 37th Meetings Held at the Palais des Nations, Geneva, from 9 to 26 March 1964' (24 August 1964) UN Doc A/AC.105/C.2/SR.29–37, 97 and 102.

Subcommittee, stating that a binding declaration that stipulates that outer space should be used for peaceful purposes only should be adopted.[154] Although both the U.S. and the USSR responded by stating that they were in favour of the use of outer space exclusively for peaceful purposes, both states also asserted that the discussion on the use of outer space exclusively for peaceful purposes could only be achieved in the framework of general and complete disarmament and therefore should be discussed in the appropriate disarmament forum; the Eighteen-Nation Committee.[155]

In COPUOS, the deliberations continued, with the prohibition of the military use of outer space being brought up once more. Austria stated that it strongly favoured the principle that outer space should be used exclusively for peaceful purposes.[156] According to the Austrian position, the absence of such a provision meant that states were still *de facto* free to conduct military uses in outer space, including the possibility to launch arms into outer space.[157] Therefore, Austria's interpretation of 'peaceful' excluded any military use and any placement of weapons into outer space. The UAR,[158] Japan,[159] India[160] and Lebanon[161] again reiterated their understanding of 'peaceful'. In its statement, Lebanon referred to the Cairo Declaration from the Second Conference of Heads of State or Government of Non-Aligned Countries of 1964, which specifically included a statement on the governance of outer space and the necessity of an international agreement prohibiting the use of outer space for military purposes.[162] Importantly, Lebanon stated that it fell upon the Legal Subcommittee to deliberate upon such an international agreement, thereby implying that the mandate of COPUOS covered military uses, or at least the prohibition thereof.

The following year, both the Legal Subcommittee and the Plenary had sessions, but most of the discussions focused on the draft agreements on assistance to and

---

[154]UNCOPUOS (Legal Subcommittee), 'Summary Records of the 29th to 37th Meetings Held at the Palais des Nations, Geneva, from 9 to 26 March 1964' (24 August 1964) UN Doc A/AC.105/C.2/SR.29–37, 106.

[155]UNCOPUOS (Legal Subcommittee), 'Summary Records of the 29th to 37th Meetings Held at the Palais des Nations, Geneva, from 9 to 26 March 1964' (24 August 1964) UN Doc A/AC.105/C.2/SR.29–37, 108.

[156]UNCOPUOS 'Verbatim Record of the Twenty-Seventh Meeting' (8 December 1964) UN Doc A/AC.105/PV.27, 14.

[157]UNCOPUOS 'Verbatim Record of the Twenty-Seventh Meeting' (8 December 1964) UN Doc A/AC.105/PV.27, 14.

[158]UNCOPUOS 'Verbatim Record of the Twenty-Eighth Meeting' (8 December 1964) UN Doc A/AC.105/PV.28, 3.

[159]UNCOPUOS 'Verbatim Record of the Twenty-Ninth Meeting' (8 December 1964) UN Doc A/AC.105/PV.29, 11.

[160]UNCOPUOS 'Verbatim Record of the Thirtieth Meeting' (8 December 1964) UN Doc A/AC.105/PV.30, 12.

[161]UNCOPUOS 'Verbatim Record of the Thirty-First Meeting' (8 December 1964) UN Doc A/AC.105/PV.31, 20.

[162]UNCOPUOS 'Verbatim Record of the Thirty-First Meeting' (8 December 1964) UN Doc A/AC.105/PV.31, 20.

return of astronauts and spacecraft and on liability for damage caused by objects launched into outer space.[163] With respect to a draft agreement on legal principles governing activities in the exploration and use of outer space, the Legal Subcommittee stated that work should start immediately, a recommendation that was adopted in the Plenary session of COPUOS.[164]

The same states that had already put forward their position on military uses of outer space reiterated their position in the First Committee.[165] With respect to the yearly UNGA resolution on international cooperation in the peaceful use of outer space, Cameroon made the suggestion to amend the resolution to include the following preambular paragraph: 'Convinced that to benefit mankind the exploration and use of outer space should be carried out solely for peaceful purposes.'[166] The U.S. supported this amendment, stating:

> The United States had constantly endorsed the principle that outer space should be used for 'peaceful purposes'. In that context, 'peaceful' meant non-aggressive rather than non-military. The United States space programme had been notable for its predominantly civilian character but military components and personnel had made indispensable contributions. There was no practical dividing-line between military and non-military uses of space: United States and Soviet astronauts had been members of their countries' armed forces; a navigation satellite could guide a warship as well as a merchant ship; communication satellites could serve military establishments as well as civilian communities. The question of military activities in space could not be divorced from the question of military activities on Earth. The test of any space activity must therefore be not whether it was military or non-military but whether it was consistent with the [UN] Charter and other obligations of international law.[167]

This statement illustrates the U.S. interpretation of the term 'peaceful'. However, the statement seems to contradict itself with respect to the U.S. position on the mandate of COPUOS. As can be read in the statement, the U.S. considers the question of military activities in space intertwined with the question of military activities on Earth, thereby reiterating their position that military activities in space should be discussed in the disarmament framework. However, stating that peaceful means non-aggressive rather than non-military also implies that COPUOS has the mandate to discuss the military but non-aggressive uses of outer space because it has the mandate to discuss the peaceful uses of outer space. This contravenes the position taken earlier by the U.S. with respect to the inclusion of provisions on war propaganda and espionage satellites in the Declaration of Legal Principles. In

---

[163] UNGA First Committee (20th Session) 'Report of the Committee on the Peaceful Uses of Outer Space' (12 October 1965) UN Doc A/6042, par. 17.

[164] UNGA First Committee (20th Session) 'Report of the Committee on the Peaceful Uses of Outer Space' (12 October 1965) UN Doc A/6042, par. 17.

[165] UNGA First Committee (20th Session) 'Summary Record of the 1421st' (18 December 1965) UN Doc A/C.1/SR.1421, 425 & 427.

[166] UNGA 'Report of the First Committee' UN GOAR 20th Session UN Doc A/6212 (20 December 1965).

[167] UNGA First Committee (20th Session) 'Summary Record of the 1422nd Meeting' (20 December 1965) UN Doc A/C.1/SR.1422, 429.

addition, the U.S. asserts that activities in outer space are inherently dual use, which means that the military and non-military uses of outer space are inherently intertwined and cannot be discussed in separate forums without any sort of collaboration. This statement indicates that the U.S. interpretation shifted more towards the USSR interpretation of the mandate of COPUOS.

Substantial progress on the Outer Space Treaty was made during the sessions in 1966. The U.S. submitted the announcement of President Lyndon B. Johnson on the need for a treaty governing the exploration of the Moon and other celestial bodies, in which it was stated that the U.S. wanted to ensure that the exploration of the Moon and other celestial bodies would be for peaceful purposes only, including a prohibition on stationing weapons of mass destruction on a celestial body, weapon tests and military manoeuvres.[168]

Two remarks should be made pertaining to this announcement. First, the announcement specifically addressed celestial bodies, without mentioning outer space. Second, the announcement was addressed to the UNGA in general and not to COPUOS specifically. This means that the treaty envisioned by President Johnson did not necessarily need to be negotiated by COPUOS. Therefore, no consequences on the U.S. position on the mandate of COPUOS can be inferred from it. Following this announcement, the USSR formally sent a request to include an item on the UNGA agenda on the conclusion of an international agreement on legal principles governing the activities of states in the exploration and conquest of the Moon and other celestial bodies.[169] This request, which stated that the 'conquest of celestial bodies should be carried out in the interests of peace and progress exclusively' and that it was necessary 'to take steps to prohibit the use of the Moon and other celestial bodies for military purposes'[170] also only addressed the Moon and other celestial bodies; no mention of outer space in general was made. The necessity for such provisions, and the absence of a reference to outer space in general, was because military use of celestial bodies (in contrast with the military use of outer space) could not be justified by national security interests and would only serve the purpose of aggression.[171] Furthermore, the USSR clarified that such a prohibition would both serve the conclusion of an agreement on general and complete disarmament and

---

[168]UNGA 'Letter Dated 9 May 1966 from The Permanent Representative of the United States of America to the United Nations Addressed to the Secretary-General' UN GOAR 21st Session UN Doc A/6327 (10 May 1966).

[169]UNGA 'Union of Soviet Socialist Republics: Request for the Inclusion of an Item in the Provisional Agenda of the Twenty-First Session' UN GOAR 21st Session UN Doc A/6341 (31 May 1966).

[170]UNGA 'Union of Soviet Socialist Republics: Request for the Inclusion of an Item in the Provisional Agenda of the Twenty-First Session' UN GOAR 21st Session UN Doc A/6341 (31 May 1966), 2.

[171]UNGA 'Union of Soviet Socialist Republics: Request for the Inclusion of an Item in the Provisional Agenda of the Twenty-First Session' UN GOAR 21st Session UN Doc A/6341 (31 May 1966), 2.

serve international cooperation in the exploration and use of outer space.[172] This rhetoric resulted in proposing the inclusion of the following principle:

> The Moon and other celestial bodies should be used by all states exclusively for peaceful purposes. No military bases or installations of any kind, including facilities for nuclear and other weapons of mass destruction of any type, should be established on the Moon or other celestial bodies.[173]

Once more, the request was addressed to the UNGA as a whole rather than COPUOS. However, the fact that both the U.S. and the USSR included military uses of outer space in their respective announcements and requests, knowing that the treaty would logically be discussed in COPUOS, means that neither was averse to including military uses of outer space in the deliberations in COPUOS or its subcommittees. The draft treaties submitted by the U.S. and the USSR support this conclusion. In the Draft Treaty Governing the Exploration of the Moon and Other Celestial Bodies submitted by the U.S., this is exemplified by the following provision:

> Celestial bodies shall be used for peaceful purposes only. All States undertake to refrain from conducting on celestial bodies any activities such as the establishment of military fortifications, the carrying out of military manoeuvres, or testing of any type of weapons. The use of military personnel, facilities or equipment for scientific research or for any other peaceful purpose shall not be prohibited.[174]

Likewise, the USSR Draft Treaty on Principles Governing the Activities of States in the Exploration and Use of Outer Space, the Moon and Other Celestial Bodies contained the provision:

> The Parties to the Treaty undertake not to place in orbit around the earth any objects carrying nuclear weapons or other weapons of mass destruction and not to station such weapons on celestial bodies or otherwise to station them in outer space. The moon and other celestial bodies shall be used exclusively for peaceful purposes by all Parties to the Treaty. The establishment of military bases and installations, the testing of weapons and the conduct of military manoeuvres on celestial bodies shall be forbidden.[175]

Therefore, both the U.S. and USSR drafts affirmed that despite both states having taken the position in earlier deliberations that the military uses of outer space should be discussed in the disarmament framework, the U.S. and USSR still included such discussion, at least in some minimal form, in the deliberations in COPUOS. The

---

[172]UNGA 'Union of Soviet Socialist Republics: Request for the Inclusion of an Item in the Provisional Agenda of the Twenty-First Session' UN GOAR 21st Session UN Doc A/6341 (31 May 1966), 2.

[173]UNGA 'Union of Soviet Socialist Republics: Request for the Inclusion of an Item in the Provisional Agenda of the Twenty-First Session' UN GOAR 21st Session UN Doc A/6341 (31 May 1966), 2.

[174]UNCOPUOS, 'Draft Treaty Governing the Exploration of the Moon and Other Celestial Bodies' (17 June 1966) UN Doc A/AC.105/32, Article 9–10.

[175]UNGA 'Letter Dated 16 June 1966 From the Permanent Representative of the Union of Soviet Socialist Republics to the United Nations Addressed to the Secretary-General' UN GOAR 21st Session UN Doc A/6352 (16 June 1966), Article IV.

provisions went one step further and even touched on a disarmament issue by stating that no nuclear weapons or other weapons of mass destruction may be stationed on celestial bodies or stationed in outer space. Therefore, the U.S. and USSR interpretations of the mandate of COPUOS seem to have shifted towards the interpretation favoured by India, Lebanon, the UAR, etc., namely that the non-arms military uses of outer space should be discussed in COPUOS and that COPUOS could at least discuss limiting the use of outer space for exclusively peaceful purposes, for example through the prohibition of placing weapons on celestial bodies or in orbit.

Subsequently, these drafts were discussed in the Legal Subcommittee. The similarity of the two drafts was mentioned, and with respect to the aforementioned draft provisions, it was specifically stated that the conclusion of an agreement that would rule out a possible arms race or territorial claim in space would help towards keeping peace in space.[176] The U.S. reiterated its position that the exploration of the Moon and other celestial bodies should be for peaceful purposes only but added that the central objective was to ensure that outer space and celestial bodies were used for peaceful purposes exclusively.[177] Unlike in the draft treaty, the U.S. did include outer space in this statement, thereby broadening the scope of the area that should be used for peaceful purposes exclusively. However, it should be borne in mind that the U.S. interpretation of 'peaceful' meant non-aggressive and not non-military. Nevertheless, the conclusion is that the U.S. included the discussion of military uses of outer space in the deliberations in COPUOS, contrary to the position it had taken previously. Likewise, the USSR clarified its draft treaty reiterating the statements made when the request was made to include the item on the UNGA agenda,[178] namely that the use of the Moon and other celestial bodies should not be used for military uses and should be in the interest of peace and progress exclusively.[179]

Considering the earlier statements made by India, it is unsurprising that India approved of the inclusion of the aforementioned provisions in the U.S. and USSR draft treaties on the use of outer space for exclusively peaceful purposes.[180] Further

---

[176]UNCOPUOS (Legal Subcommittee), 'Summary Record of the Fifty-Seventh Meeting' (20 October 1966) UN Doc A/AC.105/C.2/SR.57, 4.

[177]UNCOPUOS (Legal Subcommittee), 'Summary Record of the Fifty-Seventh Meeting' (20 October 1966) UN Doc A/AC.105/C.2/SR.57, 6.

[178]UNCOPUOS (Legal Subcommittee), 'Summary Record of the Fifty-Seventh Meeting' (20 October 1966) UN Doc A/AC.105/C.2/SR.57, 11.

[179]UNCOPUOS (Legal Subcommittee), 'Summary Record of the Fifty-Seventh Meeting' (20 October 1966) UN Doc A/AC.105/C.2/SR.57, 11.

[180]UNCOPUOS (Legal Subcommittee), 'Summary Record of the Fifty-Seventh Meeting' (20 October 1966) UN Doc A/AC.105/C.2/SR.57, 18–20.

statements made by Austria,[181] Japan,[182] Czechoslovakia[183] and Hungary[184] all indicated an implicit acknowledgment of the mandate of COPUOS to discuss the military uses of outer space, at least in the context of a treaty governing the activities of states in outer space. Hungary even went so far as to state that COPUOS was dealing with the demilitarisation of outer space,[185] which would logically fall wholly within the mandate of the disarmament framework. In a similar manner, the UAR hoped that the deliberations of the Legal Subcommittee would lead to progress on disarmament.[186] Specifically, the UAR stated:

> It was to be regretted that the two drafts provided for the non-militarization of the moon and other celestial bodies but not for that of outer space.[187]

Both Hungary and the UAR therefore implied that when it came to the use of outer space, the mandate of COPUOS extended towards disarmament matters. This sentiment was echoed by Canada, which even stated that a treaty establishing an international legal order in outer space would be a welcome addition to arms control measures.[188] Likewise, Argentina put forward that outer space and celestial bodies should be used solely for peaceful purposes, which included prohibiting the placement of nuclear weapons or weapons of mass destruction in outer space, the testing of any kind of weapons, establishing military fortifications or conducting military manoeuvres.[189] Therefore, Argentina followed an approach towards peaceful uses of outer space as non-aggressive rather than non-military as it did not touch upon other military uses of outer space, such as the use of satellites to guide military movements on Earth, but an approach that allowed military uses of outer space to be discussed in COPUOS. Similarly, Poland stated that the treaty governing the activities of states in outer space would limit the arms race, thus seeing it as a means of disarmament.[190]

---

[181] UNCOPUOS (Legal Subcommittee), 'Summary Record of the Fifty-Eight Meeting' (20 October 1966) UN Doc A/AC.105/C.2/SR.58, 4.

[182] UNCOPUOS (Legal Subcommittee), 'Summary Record of the Fifty-Eight Meeting' (20 October 1966) UN Doc A/AC.105/C.2/SR.58, 6.

[183] UNCOPUOS (Legal Subcommittee), 'Summary Record of the Fifty-Eight Meeting' (20 October 1966) UN Doc A/AC.105/C.2/SR.58, 8.

[184] UNCOPUOS (Legal Subcommittee), 'Summary Record of the Fifty-Ninth Meeting' (24 October 1966) UN Doc A/AC.105/C.2/SR.59, 2–3.

[185] UNCOPUOS (Legal Subcommittee), 'Summary Record of the Fifty-Ninth Meeting' (24 October 1966) UN Doc A/AC.105/C.2/SR.59, 2–3.

[186] UNCOPUOS (Legal Subcommittee), 'Summary Record of the Sixty-Second Meeting' (24 October 1966) UN Doc A/AC.105/C.2/SR.62, 2–3.

[187] UNCOPUOS (Legal Subcommittee), 'Summary Record of the Sixty-Second Meeting' (24 October 1966) UN Doc A/AC.105/C.2/SR.62, 4.

[188] UNCOPUOS (Legal Subcommittee), 'Summary Record of the Sixty-Second Meeting' (24 October 1966) UN Doc A/AC.105/C.2/SR.62, 4.

[189] UNCOPUOS (Legal Subcommittee), 'Summary Record of the Sixtieth Meeting' (20 October 1966) UN Doc A/AC.105/C.2/SR.60, 2–3.

[190] UNCOPUOS (Legal Subcommittee), 'Summary Record of the Sixty-Second Meeting' (24 October 1966) UN Doc A/AC.105/C.2/SR.62, 7.

Mexico even went so far as to state that it hoped the treaty would make states accustomed to general and complete disarmament, which would then first be realised in outer space and would lead to progress on general and complete disarmament on Earth.[191]

The aforementioned statements of a wide array of states, and the fact that no statements were recorded to the contrary, illustrate that disarmament and military uses of outer space were discussed in COPUOS despite the fact that states indicated that they should be discussed in the disarmament framework. Therefore, in contrast to the formal statements, the actual practical mandate of COPUOS does seem to include such matters. This is further substantiated through the discussions on the specific articles that deal with the military uses of outer space, Articles 8 and 9 of the U.S. draft and Article IV of the USSR draft. In these articles, the U.S. stipulated that restrictions should be placed on military activities on celestial bodies.[192] Likewise, the USSR stipulated a limitation of the military use of outer space, including the placement of weapons of mass destruction in outer space.[193] Argentina, Hungary and India requested the inclusion of the phrase 'outer space should be used exclusively for peaceful purposes' in the articles but without questioning or debating the mandate of COPUOS to discuss these matters.[194]

Although the Legal Subcommittee did not agree upon a complete text for the treaty governing activities of states in outer space,[195] the deliberations led to the inclusion of an article in the draft treaty that stipulated the prohibitions to not place nuclear weapons or any other kind of weapons of mass destruction in outer space, to not establish military bases and fortifications and to not test any kind of weapons or conduct military manoeuvres on celestial bodies.[196] In addition, the provision determined that celestial bodies should be used exclusively for peaceful purposes but that the use of military personnel for scientific research or any other peaceful purposes would be allowed.[197]

---

[191]UNCOPUOS (Legal Subcommittee), 'Summary Record of the Sixty-Second Meeting' (24 October 1966) UN Doc A/AC.105/C.2/SR.62, 8.

[192]UNCOPUOS (Legal Subcommittee), 'Summary Record of the Sixty-Fifth Meeting' (24 October 1966) UN Doc A/AC.105/C.2/SR.65, 9.

[193]UNCOPUOS (Legal Subcommittee), 'Summary Record of the Sixty-Fifth Meeting' (24 October 1966) UN Doc A/AC.105/C.2/SR.65, 10.

[194]UNCOPUOS (Legal Subcommittee), 'Summary Record of the Sixty-Sixth Meeting' (21 October 1966) UN Doc A/AC.105/C.2/SR.66.

[195]UNCOPUOS (Legal Subcommittee), 'Summary Record of the Seventy-Third Meeting' (19 October 1966) UN Doc A/AC.105/C.2/SR.73, 14.

[196]UNCOPUOS, 'Report of the Legal Sub-Committee on the Work of Its Fifth Session (12 July–4 August and 12–16 September 1966) to the Committee of the Peaceful Uses of Outer Space' (16 September 1966) UN Doc A/AC.105/35 Annex II, 5.

[197]UNCOPUOS, 'Report of the Legal Sub-Committee on the Work of Its Fifth Session (12 July–4 August and 12–16 September 1966) to the Committee of the Peaceful Uses of Outer Space' (16 September 1966) UN Doc A/AC.105/35 Annex II, 5.

In contrast with earlier remarks made by various states, the Legal Subcommittee thus not just discussed military uses of outer space but also deliberated on disarmament matters pertaining to outer space. Therefore, the mandate of COPUOS is wider than just discussing 'peaceful uses' of outer space, especially considering that the discussion even extended into the demilitarisation of outer space. At the very least, the deliberations in the Legal Subcommittee support the idea that the term 'peaceful' uses includes non-arms military uses of outer space, which should then be discussed in COPUOS.

The discussion continued when the deliberations on the treaty moved to the Plenary sessions of COPUOS. Once more, states affirmed the fact that military uses and disarmament matters could be discussed in COPUOS. As put forward by the U.S.: 'It has also been noted (...) that important advances can be made in arms control through the medium of this treaty.'[198] France even specifically described the agreed upon articles as adopting the non-militarisation of outer space.[199] Because there were still a number of provisions on which no agreement could be reached, neither in the Legal Subcommittee nor in the Plenary, the final report of COPUOS gave an overview of the articles on which agreement had been found and on the articles that were still contentious.[200] The fact that no agreement had been reached led the U.S. to request to include the treaty on the agenda of the 21st session of the UNGA.[201] Likewise, the USSR submitted a revised draft treaty for consideration at the 21st session, taking into consideration the deliberations and consensus reached in the Legal Subcommittee.[202]

However, before the treaty could be discussed by the First Committee, consultations between states had resolved the contentious articles, and a full draft treaty was presented to the First Committee in a draft resolution supported by 37 states (which later received support of another six states becoming the Forty-Three-Power draft resolution).[203] In the discussion on the draft treaty, the U.S. once more reiterated that it dealt with disarmament and the regulation of arms, even referring to a statement by

---

[198] UNCOPUOS, 'Verbatim Record of the Forty-Fourth Meeting' (25 October 1966) UN Doc A/AC.105/PV.44, 20.

[199] UNCOPUOS, 'Verbatim Record of the Forty-Fourth Meeting' (25 October 1966) UN Doc A/AC.105/PV.44, 39–40.

[200] UNCOPUOS, 'Report of the Committee on the Peaceful Uses of Outer Space' (22 September 1966) UN Doc A/6431.

[201] UNGA 'United States of America: Request for the Inclusion of an Additional Item in the Agenda of the Twenty-First Session' UN GOAR 21st Session UN Doc A/6392 (19 September 1966).

[202] UNGA 'Letter Dated 4 October 1966 from the Representative of the Union of Soviet Socialist Republics to the Secretary-General' UN GOAR 21st Session UN Doc A/6532/REV.1 (5 October 1966).

[203] UNGA First Committee (21st Session) 'Afghanistan, Argentina, Australia, Austria, Belgium, Brazil, Bulgaria, Canada, Chad, Czechoslovakia, Dahomey, Denmark, Finland, France, Hungary, Iran, Iraq, Ireland, Italy, Japan, Jordan, Lebanon, Mexico, Mongolia, Morocco, Nepal, Niger, Poland, Romania, Sierra Leone, Sudan, Sweden, Turkey, Union of Soviet Socialist Republics, United Kingdom of Great Britain and Northern Ireland, United States of America and Uruguay: draft resolution' (15 December 1966) UN Doc A/C.1/L.396.

President Johnson that the treaty was the 'most important arms control development since the limited test ban treaty of 1963'.[204] Likewise, the UK stated that Article IV of the treaty was of great importance to the field of disarmament,[205] while Brazil referred to Article IV as a partial disarmament measure.[206] Unsurprisingly, considering the wide support for the draft resolution, it was adopted without objection.[207] Subsequently, the treaty was unanimously adopted by the UNGA.[208] Resolution 2222 (XXI) commends the Outer Space Treaty and also stipulates the specific mandate of COPUOS within the general mandate provided at the establishment of COPUOS, namely that COPUOS should focus on agreements on liability and on the rescue and return of astronauts and study questions on the definition of outer space and the utilisation of outer space and celestial bodies.[209]

The discussions at every level of the UN framework illustrate that states had no issue with deliberating on military uses of outer space or outer space disarmament matters in the context of COPUOS. Therefore, it seems that a shift has occurred during the negotiations of the OST on the interpretation of the mandate of COPUOS. The U.S. and the USSR seem to have changed their interpretation and to have sided with the interpretation favoured by India. The conclusion can then be drawn that this gives an implicit mandate to COPUOS to discuss military uses of outer space and outer space disarmament matters. However, such a conclusion would be hasty. Indeed, the OST was largely negotiated in the Legal Subcommittee and the Plenary, and it does touch upon military uses of outer space and the disarmament of outer space as exemplified by the statements made by various states. However, the matter of preparing a treaty on governing the activities of states in outer space was first requested to be put on the agenda of the 21st Session of the UNGA. The fact that it was first requested to be put on the agenda of the UNGA and was only subsequently added to the agenda of COPUOS could be considered as COPUOS having been given a special mandate to include a discussion of military uses of outer space and the disarmament of outer space within the context of an all-encompassing treaty governing the activities of states in outer space rather than an implicit mandate to discuss any and all military uses of outer space and the disarmament of outer space at any time. The further deliberations in COPUOS following the adoption of the OST will demonstrate whether COPUOS had a special mandate to discuss military uses of

---

[204]UNGA First Committee (21st Session) 'Verbatim Record of Fourteen Hundred and Ninety-Second Meeting' (27 January 1962) UN Doc A/C.1/PV.1492, 428.

[205]UNGA First Committee (21st Session) 'Verbatim Record of Fourteen Hundred and Ninety-Second Meeting' (27 January 1962) UN Doc A/C.1/PV.1492, 430.

[206]UNGA First Committee (21st Session) 'Verbatim Record of Fourteen Hundred and Ninety-Second Meeting' (27 January 1962) UN Doc A/C.1/PV.1492, 432.

[207]UNGA First Committee (21st Session) 'Report of the First Committee' (17 December 1966) UN Doc A/6621, 26.

[208]UNGA 'Provisional Verbatim Record of the Fourteen Hundred and Ninety-Ninth Plenary Meeting' UN GOAR 21st Session UN Doc A/PV.1499 (19 December 1966), 71.

[209]UNGA Res 2222 (XXI) Treaty on Principles Governing the Activities of States in the Exploration and Use of Outer Space, including the Moon and Other Celestial Bodies (19 December 1966).

3.2.1.3 Agreement on the Rescue of Astronauts, the Return of Astronauts and the Return of Objects Launched into Outer Space (ARRA)

Simultaneous with the negotiations on the Declaration of Legal Principles and the OST, COPUOS deliberated on an international agreement on the rescue and return of astronauts and space vehicles (ARRA). The earlier deliberations on the ARRA occurred during the deliberations on Resolution 1962 (XVIII). The interpretations of the mandate of COPUOS are thus similar to the interpretations shown in the section on Resolution 1962 (XVIII). Therefore, the USSR draft international agreement on the rescue of astronauts and spaceships making emergency landings contained a provision that pertains to the peaceful and military uses of outer space, namely in Article 7:

> Space vehicles aboard which devices have been discovered for the collection of intelligence information in the territory of another State shall not be returned.[210]

This provision and the fact that the USSR wanted it included in the agreement led to discussions on the mandate of COPUOS, specifically whether that mandate extended to discussing such non-aggressive or non-arms military uses of outer space. Essentially, the gathering of intelligence information, or espionage, is the best example of a non-aggressive military use of outer space. In contrast, and similar to the discussions held with respect to the Declaration of Legal Principles, the U.S. draft proposal on assistance to and return of space vehicles and personnel did not touch upon the military uses of outer space.[211]

As was the case with the discussion on the Declaration of Legal Principles, there were divergent views on the mandate of COPUOS to discuss such issues. Czechoslovakia, for example, stated that a spacecraft that was launched for purposes incompatible with peaceful co-existence and international cooperation would not deserve the same humanitarian assistance as a spacecraft launched for peaceful purposes.[212] Canada exemplified the difference between the two drafts:

---

[210] UNGA 'Report of the Committee on the Peaceful Uses of Outer Space' UN GAOR 17th Session UN Doc A/5181 Annex I–B (27 September 1962), Article 7.

[211] UNGA 'Report of the Committee on the Peaceful Uses of Outer Space' UN GAOR 17th Session UN Doc A/5181 Annex I–C (27 September 1962).

[212] UNCOPUOS (Legal Subcommittee), 'Summary Record of the Eighth Meeting' (21 August 1962) UN Doc A/AC.105/C.2/SR.8, 6.

3 The Development of the Mandates of the Committee on the Peaceful Uses...

There appeared to be only one important difference between the drafts submitted by the United States and the Soviet Union: Article 7 of the Soviet draft required the launching State to give advance notice of each launching and placed in a special category space vehicles engaged in the collection of intelligence.[213]

Canada further stated that such issues did not fall within the mandate of COPUOS but 'within the term of reference of the Eighteen-Nation Committee'.[214] Therefore, Canada took a strong position on the question of whether such matters fell within the mandate of COPUOS. The UK also commented on Article 7, stating that such an article could not be adopted by a Legal Subcommittee and should be a top-level political decision but without asserting where such a decision should be taken, at the level of COPUOS or in the Eighteen-Nation Committee.[215] However, most of the discussion on the ARRA centred around the form of the agreement—resolution versus treaty—rather than the substance or whether COPUOS was mandated to discuss uses of outer space such as espionage. These deliberations aside, neither the Legal Subcommittee nor the Plenary made much progress on the discussion of the ARRA, with no agreement emerging on the subject.[216]

The subsequent year, the USSR draft was unchanged and still contained Article 7, stipulating that those space vehicles that contained devices used for espionage would not have to be returned.[217] However, much of the debate during the 1963 session of the Legal Subcommittee focused on the negotiations on the Declaration of Legal Principles, and as such the ARRA was discussed minimally. Likewise, much of the discussion in the Plenary session of COPUOS dealt with the Declaration of Legal Principles and resulted merely in a 'rapprochement and clarification of ideas'.[218]

Following the adoption of the Declaration of Legal Principles, the attention of COPUOS and its Legal Subcommittee shifted back to the agreements on international liability and on the rescue and return of astronauts and space vehicles. A renewed proposal on the ARRA by the U.S. was submitted, much more detailed than their previous draft.[219] Nonetheless, the U.S. did not add any provision on altering

---

[213] UNCOPUOS (Legal Subcommittee), 'Summary Record of the Ninth Meeting' (21 August 1962) UN Doc A/AC.105/C.2/SR.9, 5.

[214] UNCOPUOS (Legal Subcommittee), 'Summary Record of the Ninth Meeting' (21 August 1962) UN Doc A/AC.105/C.2/SR.9, 5.

[215] UNCOPUOS (Legal Subcommittee), 'Summary Record of the Eleventh Meeting' (21 August 1962) UN Doc A/AC.105/C.2/SR.11, 5.

[216] UNGA 'Report of the Committee on the Peaceful Uses of Outer Space' UN GAOR 17th Session UN Doc A/5181 (27 September 1962), 5.

[217] UNCOPUOS (Legal Subcommittee), 'Report of the Legal Sub-Committee on the Work of Its Second Session (16 April–3 May 1963) to the Committee on the Peaceful Uses of Outer Space' (6 May 1963) UN Doc A/AC.105/12 Annex I, 4.

[218] UNGA 'Report of the Committee on the Peaceful uses of Outer Space' UN GOAR 18th Session UN Doc A/5449 (24 September 1963), par. 19.

[219] UNCOPUOS (Legal Subcommittee), 'United States: Proposal – International Agreement on Assistance to and Return of Astronauts and Objects Launched into Outer Space' (9 March 1964) UN Doc A/AC.105/C.2/L.9.

the obligations of rescue and return with respect to space objects used for gathering intelligence information or espionage. The USSR revised draft had been toned down compared to the earlier version and had removed the explicit reference to intelligence gathering or espionage. However, the USSR stressed that it still wanted to impose an exemption on the obligation to return a space vehicle if that space vehicle was not used for the peaceful exploration of outer space.[220] The USSR draft did not specify which uses would be considered to be peaceful and which uses would not.

The inclusion once more illustrates the split stance that the USSR had on the mandate of COPUOS. On the one hand, it had made clear statements that military uses of outer space and disarmament should be discussed in the appropriate disarmament forum; on the other, it did include certain military uses of outer space in its drafts, thereby implying that COPUOS had the mandate to discuss those issues. The rephrasing of the exemption on returning space vehicles did yield results, making it more acceptable for other states.[221] However, the further proposal made by Canada and Australia based on the U.S. and USSR draft did not include an exemption on the obligation to return the space vehicle.[222] In the end, the combination of the removal of the explicit reference to the gathering of intelligence information in the USSR draft and the proposal submitted by Canada and Australia led to the discussion drifting away from the use of outer space for military purposes.

The 1965 session of the Legal Subcommittee saw renewed efforts on the conclusion of an international agreement on the assistance to and return of astronauts and spacecraft. Although the USSR did not refer to gathering intelligence information specifically, it did highlight that no agreement was found on the obligation to return foreign spacecraft, stating;

> In the opinion of the Soviet delegation and of many others, the contracting parties should be obligated under the agreement only to return ships and crews launched in accordance with the principles and objectives of the 1963 Declaration of Legal Principles.[223]

In essence, the USSR thereby proposed that spacecraft not launched for peaceful purposes and in accordance with the objective of international cooperation should not be returned. The USSR made this even more explicit in the further discussion of the draft stating that the limitation on the obligation to return a foreign spacecraft was essential because otherwise a state would be obligated to fulfil its obligation under the treaty even when a spacecraft was launched 'with intentions hostile to its own

---

[220] UNCOPUOS (Legal Subcommittee), 'Summary Records of the 29th to 37th Meetings Held at the Palais des Nations, Geneva, from 9 to 26 March 1964' (24 August 1964) UN Doc A/AC.105/C.2/SR.29–37, 33.

[221] UNCOPUOS (Legal Subcommittee), 'Summary Records of the 29th to 37th Meetings Held at the Palais des Nations, Geneva, from 9 to 26 March 1964' (24 August 1964) UN Doc A/AC.105/C.2/SR.29–37, 40.

[222] UNCOPUOS 'Report of the Legal Sub-Committee on the Work of Its Third Session (9–26 March 1964) to the Committee on the Peaceful Uses of Outer Space' (26 March 1964) A/AC.105/19/Annex I, 11–12.

[223] UNCOPUOS (Legal Subcommittee), 'Summary Record of the Forty-First Meeting' (30 November 1965) UN Doc A/AC.105/C.2/SR.41, 4.

interests'.[224] The USSR saw the limitation as necessary because otherwise a state would be required to return even an unexploded shell to the country that had fired it.[225] The UK, Austria and the U.S., however, pointed out that the USSR principle would give room to subjective interpretations of one state on whether a spacecraft was launched in accordance with the Declaration of Legal Principles.[226]

This same issue was present in the discussion of the obligation to return astronauts. The USSR stated that it thought a state could not be obligated to return crews of spacecraft launched for hostile purposes.[227] The USSR position was elaborated by Bulgaria's statement that astronauts conducting military activities could not be considered 'envoys of mankind'.[228] Nonetheless, the U.S. maintained its position that the return of such a crew should be unconditional.[229] Likewise, France stated that military personnel should not receive any special advantages but should also not be excluded from getting the assistance that others would be given.[230] In the end, no agreement was reached on an international agreement on the rescue and return of astronauts and space vehicles.[231] The deliberations show, however, that states had no objections to discuss at least certain aspects of the military use of outer space in COPUOS.

Once more, the discussion on the ARRA was interrupted by deliberations on a different legal instrument (the OST) in the Legal Subcommittee and the Plenary session of COPUOS. As stated in the section on the OST, the deliberations pertaining to that instrument included a number of provisions that touched upon the use of outer space for military purposes and thus the mandate of COPUOS to discuss such issues. The further deliberations after the adoption of the OST will illustrate whether this was deliberate, and has ramifications for the mandate of COPUOS, or such matters were merely discussed in the context of the OST to have an all-encompassing treaty governing activities in outer space. In addition, Resolution 2222 (XXI) had specified the mandate of COPUOS for the subsequent

---

[224]UNCOPUOS (Legal Subcommittee), 'Summary Record of the Forty-Second Meeting' (30 November 1965) UN Doc A/AC.105/C.2/SR.42, 5.

[225]UNCOPUOS (Legal Subcommittee), 'Summary Record of the Forty-Second Meeting' (30 November 1965) UN Doc A/AC.105/C.2/SR.42, 13.

[226]UNCOPUOS (Legal Subcommittee), 'Summary Record of the Forty-Second Meeting' (30 November 1965) UN Doc A/AC.105/C.2/SR.42, 8–9, 9, 13.

[227]UNCOPUOS (Legal Subcommittee), 'Summary Record of the Forty-Sixth Meeting' (30 November 1965) UN Doc A/AC.105/C.2/SR.46, 3.

[228]UNCOPUOS (Legal Subcommittee), 'Summary Record of the Forty-Sixth Meeting' (30 November 1965) UN Doc A/AC.105/C.2/SR.46, 4.

[229]UNCOPUOS (Legal Subcommittee), 'Summary Record of the Forty-Sixth Meeting' (30 November 1965) UN Doc A/AC.105/C.2/SR.46, 3.

[230]UNCOPUOS (Legal Subcommittee), 'Summary Record of the Forty-Seventh Meeting' (30 November 1965) UN Doc A/AC.105/C.2/SR.47, 9.

[231]UNCOPUOS 'Report of the Legal Sub-Committee on the Work of Its Fourth Session (20 September–1 October 1965) to the Committee on the Peaceful Uses of Outer Space' (1 October 1965) UN Doc A/AC.105/29, 2.

year, namely that it should work on the international agreements on liability of states and on the rescue and return of astronauts and spacecraft and study the question of the definition of outer space and the utilisation of outer space and celestial bodies.[232]

With respect to the agreement on the return and rescue of astronauts and spacecraft, the USSR submitted a revised draft that removed any mention of the return of astronauts and spacecraft altogether and merely focused on the rescue of astronauts,[233] while the revised Australian/Canadian proposal did include the return of astronauts and spacecraft but made no exemption on the obligation to return spacecraft.[234] The removal of the obligation to return astronauts and spacecraft led to a lot of time being spent in the Legal Subcommittee on discussing the scope of the agreement.[235] Differences on the scope of the agreement made it impossible for COPUOS to come to an agreement. Instead, COPUOS merely expressed the hope that the Legal Subcommittee would be able to achieve more in its next session.[236] This shift in the deliberations resulted in fewer relevant statements. Occasionally, however, statements were made on the issue, such as that of the Netherlands in the First Committee:

> At the same time, the [Outer Space Treaty] is not a model of perfection and its shortcomings point to the moral that international instruments of this kind can rarely reverse the course of events which have already taken place. The best one can do, as is exemplified by the outer space treaty, is to reconcile the 'ideal' with the prevailing 'hard facts' of the political and military situation.

This statement does not directly refer to the perceived mandate of COPUOS. Rather, it acknowledges the *status quo* of the use of outer space for military purposes, namely that such uses were already part of the exploration and use of outer space and that it would take further agreements to completely prohibit such use of outer space. Nevertheless, the statement also acknowledges that the OST did touch upon issues relating to the military use of outer space, which is exemplified by the sentence 'to reconcile the 'ideal' with the prevailing 'hard facts' of the political and military situation', without criticising or commenting on a COPUOS-negotiated treaty stipulating such provisions. Therefore, and as described in the preceding, there

---

[232]UNGA Res 2222(XXI) Treaty on Principles Governing the Activities of States in the Exploration and Use of Outer Space, Including the Moon and Other Celestial Bodies (19 December 1966).

[233]UNCOPUOS (Legal Subcommittee), 'USSR: Revised Draft – Agreement on the Rescue of Astronauts in the Event of Accident or Emergency Landing' (19 June 1967) UN Doc A/AC.105/C.2/L.18 I UNCOPUOS (Legal Subcommittee), 'Summary Record of the Seventy-Sixth Meeting' (13 November 1967) UN Doc A/AC.105/C.2/SR.76, 5.

[234]UNCOPUOS (Legal Subcommittee), 'Assistance to and Return of Astronauts and Space Objects Revised Australian – Canadian Proposal Submitted as a Working Paper' (21 June 1967) UN Doc A/AC.105/C.2/L.20.

[235]UNCOPUOS 'Provisional Verbatim Record of the Forty-Ninth Meeting' (13 September 1967) UN Doc A/AC.105/PV.49, 36.

[236]UNGA 'Report of the Committee on the Peaceful Uses of Outer Space' UN GOAR 22nd Session UN Doc A/6804 (27 September 1967).

was an implied mandate for COPUOS to discuss such issues, at least when it was considering the OST.

The stalemate on the ARRA in the Legal Subcommittee permeated into the Plenary and the First Committee. The discussion in the First Committee led to the adoption of Resolution 2260(XXII) by the UNGA, which urged COPUOS to continue its work on the ARRA and the agreement on liability 'with a sense of urgency'.[237] This resolution led to informal consultations in which further reconciliation was found between the U.S. and the USSR on the provisions that should be included in the ARRA.[238] In turn, this led to a working paper consisting of a draft agreement on the rescue of astronauts, the return of astronauts and the return of objects launched into outer space.[239] Unlike the previous USSR drafts, this working paper did not include a reference to intelligence information or even to the need for a spacecraft to be launched for peaceful purposes or in accordance with the Declaration of Legal Principles or the OST.

However, this did not prevent other states from commenting on certain military issues. As the preamble of the draft agreement referred to the peaceful exploration and use of outer space, Japan stated that such a mention should not be hollow and that the ARRA could not obligate the return of space objects used in violation of Article IV OST.[240] The statement did not lead to any responses by other states and was not mentioned in the report of the Legal Subcommittee in which the draft agreement was presented to COPUOS.[241] In the end, the ARRA was adopted by the UNGA without any mention of any use of outer space for military purposes other than the preambular paragraph 'wishing to promote international co-operation in the peaceful exploration and use of outer space'.[242]

---

[237] UNGA Res 2260(XXII) 'Report of the Committee on the Peaceful Uses of Outer Space' (3 November 1967) I UNCOPUOS (Legal Subcommittee), 'Summary Record of the Eighty-Sixth Meeting' (9 February 1968) UN Doc A/AC.105/C.2/SR.86, 3.

[238] UNCOPUOS (Legal Subcommittee), 'Summary Record of the Eighty-Sixth Meeting' (9 February 1968) UN Doc A/AC.105/C.2/SR.86, 3.

[239] UNCOPUOS (Legal Subcommittee), 'Draft Agreement on the Rescue of Astronauts, the Return of Astronauts, and the Return of Objects Launched into Outer Space' (12 December 1967) UN Doc A/AC.105/C.2/L.28.

[240] UNCOPUOS (Legal Subcommittee), 'Summary Record of the Eighty-Sixth Meeting' (9 February 1968) UN Doc A/AC.105/C.2/SR.86, 11.

[241] UNCOPUOS 'Report of the Legal Sub-Committee on the Work of Its Special Session (14–15 December 1967) to the Committee to the Peaceful Uses of Outer Space' (15 December 1967) UN Doc A/AC.105/43.

[242] UNGA Res 2345 (XXII) Agreement on the Rescue of Astronauts, the Return of Astronauts and the Return of Objects Launched into Outer Space (19 December 1967).

3.2.1.4 Convention on International Liability for Damage Caused by Space Objects (LIAB)

In the same period that the Declaration of Legal Principles, the OST and the ARRA were discussed, negotiated and adopted, an agreement on the liability of states for damage caused by objects launched into outer space was on the agenda of COPUOS. This agreement was eventually adopted as the Convention on International Liability for Damage Caused by Space Objects (LIAB). For the purposes of this report, *i.e.* on the collaboration between and mandate of COPUOS and the CD, the LIAB is of limited interest. Unlike the aforementioned instruments, even in its earliest drafts[243] the LIAB does not touch upon the use of outer space for military purposes. Although the LIAB is also applicable to military satellites and military space vehicles causing damage, it did not lead to any discussion on the mandate of COPUOS, nor did it otherwise have any implications on that mandate. Moreover, because the negotiations happened largely at the same time as the negotiations of the other instruments, the statements and positions on the mandate of COPUOS and the developments thereon have been sufficiently illustrated.

3.2.1.5 Convention on Registration of Objects Launched in Outer Space (REG)

The Convention on Registration of Objects Launched in Outer Space (REG), like the LIAB, does not elucidate the mandate of COPUOS or its interpretation by its member states. Other than an acknowledgement in the preamble of the drafts of the REG, submitted by France,[244] Canada[245] and the U.S.,[246] there is no reference to the peaceful use of outer space or to any military aspect of the use of outer space. None of the provisions in the draft attempted to make an exception to the obligation to register military space objects launched into outer space.

Moreover, the drafts also did not specifically mention the registration of military space objects. This is exemplified in the U.S. draft, which was the most precise in stipulating the information that needs to be furnished during registration. In Article IV of that draft, it is stipulated that the general function of the space object needs to be furnished upon registration. This is similar to the other drafts except that it

---

[243]UNCOPUOS 'United States of America: Draft Proposal on Liability for Space Vehicle Accidents' (11 September 1962) UN Doc A/AC.105/L.5 | UNCOPUOS (Legal Subcommittee), 'Hungary Proposed Draft Agreement: Agreement Concerning Liability for Damage Caused by the Launching of Objects into Outer Space' (16 March 1964) UN Doc A/AC.105/C.2/L.10.

[244]UNCOPUOS (Legal Subcommittee), 'France: Proposal – Draft Convention Concerning the Registration of Objects Launched into Space for the Exploration of Use of Outer Space' (18 June 1968) UN Doc A/AC.105/C.2/L.45.

[245]UNCOPUOS (Legal Subcommittee), 'Canada: Draft Convention on Registration of Objects Launched into Outer Space' (4 April 1972) UN Doc A/AC.105/C.2/L.82.

[246]UNCOPUOS (Legal Subcommittee), 'United States of America: Proposal – Convention on the Registration of Objects Launched into Outer Space' (19 March 1973) UN Doc A/AC.105/C.2/L.85.

exhaustively specifies the four possible categories of general function of the space object: development of space flight techniques and technology, space research and exploration, practical applications of space based on technology and non-functional objects.[247] The military use of outer space could fall under all of these four categories, but no explicit mention is made of a military function in general. It is therefore not surprising that the discussions on the REG did not pertain to the mandate of COPUOS or the related issue of the peaceful or military use of outer space or that the final version of the REG did not have any bearing on this matter either.[248] The deliberations on this treaty thus give no insight into the development of the mandate of COPUOS or the development of the interpretation of the mandate by the member states of COPUOS.

3.2.1.6 Agreement Governing the Activities of States on the Moon and Other Celestial Bodies (MOON)

The deliberations on the Agreement Governing the Activities of States on the Moon and Other Celestial Bodies (MOON) began as an extension and development of the principles contained in the OST, analogous to the ARRA, LIAB and REG. In 1966, even before the adoption of the OST, the U.S. had proposed a treaty concerning the governance of the use of the Moon.[249] The U.S. proposal regarded the 'use for peaceful purposes only' as an essential element of a treaty governing the use of the Moon, as well as the prohibition of the placement of weapons of mass destruction, the establishment of military fortifications, the carrying out of weapons tests and the carrying out of military manoeuvres.[250] However, other deliberations in COPUOS became more prominent (OST, ARRA, LIAB, REG), and the U.S. proposal shifted to the background.

In 1969, a renewed proposal came from Poland, elaborating the OST principles with respect to activities on the Moon and other celestial bodies. The Polish draft was followed by a similar proposal from Argentina (which made a proposal relating to 'the legal status of substances, resources and products originating from the

---

[247] UNCOPUOS (Legal Subcommittee), 'United States of America: Proposal – Convention on the Registration of Objects Launched into Outer Space' (19 March 1973) UN Doc A/AC.105/C.2/L.85, 3.

[248] Convention on Registration of Objects Launched into Outer Space (adopted 12 November 1974, entered into force 15 September 1976) 1023 UNTS 15.

[249] UNCOPUOS 'Letter dated 16 June 1966 from the Permanent Representative of the United States of America addressed to the Chairman of the Committee on the Peaceful Uses of Outer Space' (17 June 1966) UN Doc A/AC.105/32.

[250] UNCOPUOS 'Letter dated 16 June 1966 from the Permanent Representative of the United States of America addressed to the Chairman of the Committee on the Peaceful Uses of Outer Space' (17 June 1966) UN Doc A/AC.105/32, 2.

moon').[251] These proposals were combined, and with the addition of France, a proposal was submitted to include on the agenda of the Legal Subcommittee 'questions relating to the legal rules which should govern man's activities on the moon and other celestial bodies, including the legal regime governing substances coming from the moon and from other celestial bodies'.[252] In 1970, Argentina once more submitted a proposal focused on governing the use of the natural resources of the Moon and other celestial bodies.[253] However, most of the focus during the ninth session of COPUOS was on the LIAB, with the MOON receiving little attention. Furthermore, the Argentinian proposal was fully focused on the use of natural resources rather than matters related to the military use of outer space.

Instead, these issues once more came into play from the moment the USSR submitted its draft treaty concerning the Moon.[254] Most of the draft is focused on the specific governance of the use of the Moon and other celestial bodies compared to the general governance contained in the OST. The USSR draft contains provisions on, *inter alia,* the Moon's environment, non-appropriation of the Moon's natural resources and furnishing information on missions to the Moon. However, the draft also specifically deals with the military use of the Moon. First, Article I reiterates Article III OST on the applicability of international law and specifically the UN Charter to the activities conducted on the Moon and in orbit around the Moon. Moreover, it includes a prohibition on the threat or use of force (Art. 2(4) UN Charter) explicitly applicable to the Moon, also in relation to the Earth or space objects.[255] Second, Article II reiterates the principles found in Article IV OST, namely that the Moon shall be used exclusively for peaceful purposes, the prohibition of the placement of weapons of mass destruction into orbit and the prohibition on establishing any military bases.[256]

The drafts submitted by the U.S. in 1966 and by the USSR in 1971 once more illustrate the willingness of both states to discuss the military use of outer space in COPUOS or at least the limitation of the military use of outer space. The discussion of this topic was supported by 10 other states, which in their draft resolution

---

[251] UNCOPUOS 'Report of the Legal Sub-committee on the Work of its Eighth Session (9 June–4 July 1969) to the Committee on the Peaceful Uses of Outer Space' (4 July 1969) UN Doc A/AC.105/58, 3.

[252] UNCOPUOS 'Report of the Legal Sub-committee on the Work of its Eighth Session (9 June–4 July 1969) to the Committee on the Peaceful Uses of Outer Space' (4 July 1969) UN Doc A/AC.105/58, 5.

[253] UNCOPUOS 'Report of the Legal Sub-committee on the Work of its Ninth Session (8 June–3 July 1970) to the Committee on the Peaceful Uses of Outer Space' (3 July 1970) UN Doc A/AC.105/85 Annex II, 1–2.

[254] UNGA First Committee (26th Session) 'USSR: Draft Treaty Concerning the Moon' (5 November 1971) UN Doc A/C.1/L.568.

[255] UNGA First Committee (26th Session) 'USSR: Draft Treaty Concerning the Moon' (5 November 1971) UN Doc A/C.1/L.568, 1–2.

[256] UNGA First Committee (26th Session) 'USSR: Draft Treaty Concerning the Moon' (5 November 1971) UN Doc A/C.1/L.568, 2.

emphasised the need to use the Moon exclusively for peaceful purposes and 'to prevent the Moon of becoming a scene of international conflict'.[257] This draft resolution was adopted and became UNGA Resolution 2779 (XXVI), which called for the consideration of a draft treaty concerning the Moon 'as a matter of priority'.[258]

In the discussions in the Legal Subcommittee following the USSR draft, the USSR emphasised the importance of the two aforementioned provisions, stating that the use of the Moon should be for only 'purely peaceful purposes, for the good of all mankind'.[259] Furthermore, various delegations made statements in support of limiting the military uses of outer space. Egypt and Lebanon, for example, discussed a prohibition of the placement of *all* weapons on the Moon rather than just weapons of mass destruction.[260] Likewise, the UK commended COPUOS for its efforts and success in limiting the placement of nuclear weapons in outer space.[261] Moreover, the UK discussed the proposed prohibition on the threat or use of force under Article I of the USSR draft. Such a prohibition seemed redundant to the UK because the UN Charter was already deemed applicable to the exploration and use of outer space.

The USSR draft and the subsequent statements thus illustrate the tacit agreement that COPUOS is the appropriate forum to discuss these issues. This is further exemplified by the U.S. proposals submitted in the Working Group on the MOON that discuss Article I and Article II of the USSR draft and by the fact that the Working Group kept most of the proposed Article I and Article II intact (as Article II and Article III, respectively).[262] The early deliberations on the MOON thus support the conclusion drawn with respect to the negotiations of the OST: that COPUOS is mandated to discuss the (limitation of the) military use of outer space. However, the majority of the discussion on the USSR draft and the subsequent Working Group draft was focused on the more contentious issues of the exploitation of the natural resources on the Moon and sharing the benefit of that exploitation and whether the

---

[257]UNGA First Committee (26th Session) 'Austria, Belgium, Bulgaria, Czechoslovakia, Hungary, India, Mongolia, Poland, Romania, Sweden and the Union of Soviet Socialist Republics: draft resolution' (8 November 1971) UN Doc A/C.1/L.572.

[258]UNGA Res 2779 (XXVI) 'Preparation of an International Treaty Concerning the Moon' (29 November 1971) UN Doc A/RES/2779 (XXIV).

[259]UNCOPUOS (Legal Subcommittee), 'Summary Records of the One Hundred and Eighty-Seventh to the One Hundred Ninety-First Meetings' (12 July 1972) UN Doc A/AC.105/C.2/SR.187–191, 8.

[260]UNCOPUOS (Legal Subcommittee), 'Summary Records of the One Hundred and Eighty-Seventh to the One Hundred Ninety-First Meetings' (12 July 1972) UN Doc A/AC.105/C.2/SR.187–191, 19 and 21.

[261]UNCOPUOS (Legal Subcommittee), 'Summary Records of the One Hundred and Eighty-Seventh to the One Hundred Ninety-First Meetings' (12 July 1972) UN Doc A/AC.105/C.2/SR.187–191, 23.

[262]UNCOPUOS 'Report of the Legal Sub-committee on the Work of its Eleventh Session' (10 April–5 May 1972) UN Doc A/AC.105/101, 5–8.

treaty should be applicable only to the Moon or to other celestial bodies as well.[263] The discussion on these issues continued into the First Committee.[264] The continuing difference of opinion on these issues resulted in Resolution 2915 (XXVII) calling on COPUOS to continue its deliberations on the MOON.[265]

These two elements are the main reasons why a large part of the negotiations on the MOON do not give much input on the interpretation on the mandate of COPUOS: first, because the provisions included in the draft were a repetition of the provisions already included in the OST. Therefore, states were not opposed to the inclusion of those provisions as they were merely a repetition, reinforcement and extension of already existing international obligations. The only exception was the explicit applicability of the prohibition on the threat or use of force. However, that prohibition is already included in Article 2(4) UN Charter and therefore already applicable to outer space through Article III OST. Therefore, the inclusion of that provision did not lead to a lot of discussion other than whether it was necessary to make it explicit or whether the reference to the UN Charter was sufficient. This is exemplified by the Bulgarian draft treaty that included these provisions and even extends the prohibition on the threat or use of force to any hostile act.[266]

Second, states were much more focused on the more contentious articles in the draft on benefit sharing in the exploitation of natural resources, on the scope of the treaty and on furnishing information before Moon missions. These provisions were new compared to the OST and had much farther-reaching consequences for states. In particular, the provision on benefit sharing, which is still heavily discussed today, is regarded as the key provision that contributed to states not signing and ratifying the MOON.

The focus of COPUOS during the negotiations of the MOON is exemplified by statements made by the chairman. In 1977, the chairman stated: 'The Group had decided to give priority to the question of natural resources, since many delegations felt that, if that issue were resolved, it would be easier to reach agreement on the two remaining issues, namely the scope of the treaty and the information to be furnished on missions to the moon.'[267] Likewise, in 1978, the chairman stated:

---

[263] UNCOPUOS 'Verbatim Record of the One Hundred and Tenth Meeting' (5 September 1972) UN Doc A/AC.105/PV.110 | UNCOPUOS 'Verbatim Record of the One Hundred and Twelfth Meeting' (6 September 1972) UN Doc A/AC.105/PV.112 | UNCOPUOS 'Verbatim Record of the One Hundred and Thirteenth Meeting' (7 September 1972) UN Doc A/AC.105/PV.113 | UNCOPUOS 'Verbatim Record of the One Hundred and Fifteenth Meeting' (11 September 1972) UN Doc A/AC.105/PV.115.

[264] UNGA First Committee (27th Session) 'Provisional Verbatim Record of the Eighteen Hundred and Sixty-First Meeting'-'Provisional Verbatim Record of the Eighteen Hundred and Seventy-First Meeting' (12 October–25 October 1972) UN Doc A/C.1/PV.1861-PV.1871.

[265] UNGA Res 2915 (XXVII) (9 November 1972) UN Doc A/RES/2915 (XXVII).

[266] UNCOPUOS (Legal Subcommittee), 'Draft Treaty Relating to the Moon – Bulgaria: Working Paper' (8 May 1974) UN Doc A/AC.105/C.2/L.93.

[267] UNCOPUOS (Legal Subcommittee), 'Summary Record of the 267th Meeting' (18 March 1977) UN Doc A/AC.105/C.2/SR.267, 2.

3 The Development of the Mandates of the Committee on the Peaceful Uses...

Table 3.2 Interpretation of the mandate of COPUOS in 1979

Armed military uses	Disarmament framework (but limiting the use for exclusively peaceful purposes can be done in COPUOS)
Non-arms military uses	COPUOS
Peaceful uses	COPUOS

At its eleventh session in 1972, the Sub-Committee had approved the text of a preamble and 21 articles of the draft treaty, including its final clauses. (...) At its twelfth sessions the following year, the Sub-Committee had taken note of the text of six provisions approved by the Working Group. (...) Notwithstanding those substantial achievements, the Sub-Committee had been unable to reach agreement on three main issues which still remained unresolved: the scope of the treaty: the information to be furnished on missions to the moon; and the natural resources of the moon.[268]

These two statements clearly show that the focus of the negotiations of the MOON lay not with the mandate of COPUOS or the provisions on the limitation of military uses of outer space. The second statement in particular illustrates that states had found some measure of agreement on the other provisions in the draft and that, since the twelfth session of the Legal Subcommittee in 1973, the three remaining issues had been the focus of discussion.

In conclusion, the negotiations on the MOON do not show a development in the interpretation of the mandate of COPUOS or of the interpretation of that mandate by specific member states of COPUOS. Instead, and similar to the OST, the negotiations illustrate that the member states of COPUOS did not see the mandate of COPUOS as an obstacle to the inclusion of certain military aspects of the use of outer space (or rather limiting the military use of outer space). In comparison to the diverging interpretations on the mandate at the start of the negotiations of the OST, the interpretations during the negotiations on the MOON appear to be more harmonised and confirm the conclusion drawn after the adoption of the OST, i.e. that COPUOS is the appropriate forum to discuss not only non-military uses of outer space but also those military uses of outer space that are non-aggressive and even the topic of limiting the use of outer space to ensure the exclusive peaceful use thereof. In essence, the mandate corresponds to the interpretation given by India (and a number of other states) in the negotiations of the Declaration of Legal Principles, as displayed in Table 3.2.

The adoption of the MOON,[269] through Resolution 36/68,[270] signalled the end of negotiations and deliberations on legally binding treaties in COPUOS. It also coincided with the renewed focus on the prevention of an arms race in outer space

---

[268] UNCOPUOS (Legal Subcommittee), 'Summary Record of the 285th Meeting' (15 March 1978) UN Doc A/AC.105/C.2/SR.285, 2.

[269] Agreement Governing the Activities of States on the Moon and Other Celestial Bodies (adopted 18 December 1979, entered into force 11 July 1984) 1363 UNTS 3.

[270] UNGA Res 35/68 'Agreement Governing the Activities of States on the Moon and Other Celestial Bodies' (5 December 1979) UN Doc A/RES/34/68.

in the disarmament framework through the adoption of Resolution S-10/2.[271] The deliberations on the five UN Space Treaties, and the OST and ARRA in particular, support the aforementioned interpretation of the mandate. However, the new and clear mandate, stipulated in Resolution S-10/2, for the CD to discuss the prevention of an arms race in outer space might have consequences for the interpretation of the mandate of COPUOS. A clearer mandate to discuss outer space disarmament matters in the CD might result in states altering their interpretation of the mandate of COPUOS because another forum has gained competency to deliberate on space matters. The following paragraph will discuss the major resolutions adopted by COPUOS to determine whether such a change in the interpretation of the mandate occurred.

3.2.2 The Development of the Mandate of COPUOS After the Five Space Treaties

With the adoption of the MOON, a new era in the development of international space law started, with a shift from binding international treaties to soft law resolutions.[272] These resolutions were still the result of extensive deliberations and negotiations in COPUOS and therefore still developed the mandate of COPUOS. In addition, the mandate of COPUOS has been developed through the yearly resolutions on the International Cooperation in the Peaceful Uses of Outer Space. These yearly resolutions emphasise, reaffirm and note the concerns, interests and matters of importance to the UNGA and thus the international community. These concerns, interests and matters can be expressed through the preambular paragraphs, such as those that recognise the importance of the prevention of an arms race in outer space. They can also be expressed through specific recommendations, such as stating which items COPUOS should consider in its next session. This paragraph will discuss the yearly resolutions on the International Cooperation in the Peaceful Uses of Outer Space, and some of the major resolutions adopted that developed the mandate of COPUOS with respect to the issue of the discussion of military uses.

---

[271] UNGA Res S-10/2 'Final Document of the Tenth Special Session of the General Assembly' (28 June 1978) UN Doc A/RES/S-10/2, par. 80 | UNGA 'Report of the Disarmament Commission' UN GAOR 36th Session Supp 42 UN Doc A/36/42 (2 July 1981), par. 19.

[272] Frans von der Dunk, 'International Space Law' in Frans von der Dunk and Fabio Tronchetti (eds) *Handbook of Space Law* (Edward Elgar Publishing 2015), 37–43.

3.2.2.1 Resolution 37/92: Principles Governing the Use by States of Artificial Earth Satellites for International Direct Television Broadcasting (DBS Principles)

Concurrent with the negotiations on the MOON, COPUOS started deliberating on the need for an international treaty governing the use of direct broadcasting satellites. One of the earliest drafts of such a treaty was submitted by the USSR in 1972.[273] This proposal included two provisions that related to the use of outer space for military purposes. First, it required the consent of the receiving state before television could be broadcasted to that state, which became the most controversial issue during the negotiations because of the conflict between state sovereignty and the freedom of information.[274] Second, the USSR proposal tried to regulate the content that could be broadcasted, prohibiting the broadcasting of television that would show 'war, militarism, Nazism and racial hatred'.[275] In a sense, these provisions were the continuation of the earlier deliberations on the use of satellites for war propaganda discussed with respect to Resolution 1962 (XVIII) and the OST because they tried to prohibit the use of direct television broadcasting for propaganda purposes.

In the same year as the USSR draft, the UNGA adopted Resolution 2916 (XXVII), which specifically requested COPUOS to work on legal principles governing the use of satellites for direct television broadcasting.[276] Similar to the USSR proposal, Resolution 2916 (XXVII) was 'mindful of the need to prevent the conversion of direct television broadcasting into a source of international conflict and of aggravation of the relations among States and to protect the sovereignty of States from any external interference'.[277]

Although the USSR proposal did not lead to anything concrete, it was discussed in the Working Group on Direct Broadcasting Satellites.[278] This Working Group focused its initial efforts on five subjects, namely the applicability of international law, rights and benefits of states, international cooperation, state responsibility and the peaceful settlement of disputes.[279] The discussion of the work of the Working Group in the Legal Subcommittee saw several statements on the military use of outer space. India, for example, stated that broadcasting should not be used for harmful

---

[273] Howard Anawalt, 'Direct television Broadcasting and the Quest for Communication Equality' (1984) 5 Michigan Journal of International Law 361, 363.

[274] Howard Anawalt, 'Direct television Broadcasting and the Quest for Communication Equality' (1984) 5 Michigan Journal of International Law 361, 363–365.

[275] Howard Anawalt, 'Direct television Broadcasting and the Quest for Communication Equality' (1984) 5 Michigan Journal of International Law 361, 363.

[276] UNGA Res 2916 (XXVII) (9 November 1972) UN Doc A/RES/2916(XXVII).

[277] UNGA Res 2916 (XXVII) (9 November 1972) UN Doc A/RES/2916(XXVII), preamble.

[278] UNCOPUOS, 'Report of the Legal Sub-committee on the Work of Its Thirteenth Session (6 May–31 May 1974)' (6 June 1975) UN Doc A/AC.105/133, 12.

[279] UNCOPUOS, 'Report of the Legal Sub-committee on the Work of Its Thirteenth Session (6 May–31 May 1974)' (6 June 1975) UN Doc A/AC.105/133, 13.

propaganda.[280] Czechoslovakia echoed this sentiment, connecting such a prohibition with the principle of non-intervention in the internal affairs of other states, and submitted that the use of direct broadcasting should be coherent with the aim of the UN to maintain international peace and security.[281]

While agreement was relatively easily found on other issues, such as state responsibility and the peaceful settlement of disputes, the issues of prior consent, state sovereignty and controlling of programme content remained the major focus of the deliberations on the DBS principles and formed the barrier in achieving consensus.[282] The connection between these issues and the use of outer space for military purposes was accurately stated by Chile:

> Radio broadcasting had made it possible to transmit programmes designed to interfere politically in the affairs of other countries, and direct television broadcasting by satellite could be an even more powerful weapon for hostile propaganda. Consequently, broadcasts which did not comply with the prior consent requirement should be considered unlawful and inadmissible under international law.[283]

This position was supported by Venezuela, which stressed the importance of state sovereignty, and identified the danger that unilateral broadcasting could lead to 'autocratic control of information' and 'the use of information for political and other purposes' by those states that had broadcasting capabilities.[284] This position was also generally held by the USSR and the developing states because they perceived the unfettered broadcasting of information into their state as impinging on their sovereignty and violating the principles of the non-interference in the international affairs of their state.[285]

The opposite position, held by mostly Western states, held that no prior consent was necessary because Article 19 of the Universal Declaration of Human Rights stipulates the principle of the freedom of expression and information for the individual regardless of frontiers.[286] Therefore, the prior consent requirement was

---

[280] UNCOPUOS (Legal Subcommittee), 'Summary Records of the Two Hundred and Eighth to Two Hundred and Twenty-Fifth Meetings' (4 October 1974) UN Doc A/AC.105/C.2/SR.208–225, 53.

[281] UNCOPUOS (Legal Subcommittee), 'Summary Records of the Two Hundred and Eighth to Two Hundred and Twenty-Fifth Meetings' (4 October 1974) UN Doc A/AC.105/C.2/SR.208–225, 54–57.

[282] UNCOPUOS, 'Report of the Legal Sub-committee on the Work of its Fifteenth Session (3–28 May 1976)' (28 May 1976) UN Doc A/AC.105/171 Annex II | UNCOPUOS, 'Report of the Legal Sub-committee on the Work of its Sixteenth Session (14 March–8 April 1977)' (11 April 1977) UN Doc A/AC.105/196 Annex II | UNGA 'Report of the Committee on the Peaceful Uses of Outer Space' UN GOAR 33rd Session Supp No. 20 UN Doc A/33/20 (7 August 1978), 10–11.

[283] UNCOPUOS (Legal Subcommittee), 'Summary Record of the 269th Meeting' (22 March 1977) UN Doc A/AC.105/C.2/SR.269, 2–3.

[284] UNCOPUOS (Legal Subcommittee), 'Summary Record of the 269th Meeting' (22 March 1977) UN Doc A/AC.105/C.2/SR.269, 5.

[285] UNCOPUOS (Legal Subcommittee), 'Summary Record of the 288th Meeting' (20 March 1978) UN Doc A/AC.105/C.2/SR.288, 9.

[286] UNCOPUOS (Legal Subcommittee), 'Summary Record of the 269th Meeting' (22 March 1977) UN Doc A/AC.105/C.2/SR.269, 5–9 | UNCOPUOS (Legal Subcommittee), 'Summary Record of the 288th Meeting' (20 March 1978) UN Doc A/AC.105/C.2/SR.288, 10.

perceived as a violation of the freedom of information because it restricted that freedom, or as put by the Netherlands:

> The paramount importance of the universally recognized right of everyone to seek, receive and impart information and ideas regardless of frontiers implied that international regulation imposed no restrictions on such freedom other than those strictly required by technical constraints. (...) it was possible to intensify international co-operation through appropriate notification and consultations in order to prevent or settle disputes which might arise from the lawful or unlawful behaviour of a particular State. But freedom of information, regardless of frontiers, should remain the governing principle.[287]

Although the discussion on state sovereignty and the freedom of information is not directly related to the use of outer space for military purposes, the statements illustrate how direct broadcasting can be used to achieve certain military goals and how many states perceived the issues as related. The inclusion of these issues in the deliberations on the DBS principles was not called into question by other states, nor were there any statements that COPUOS was not the appropriate forum to discuss such matters, which corresponds with the general trend observed during the negotiations of the MOON.

This division remained the stumbling block in achieving consensus in COPOUS on the DBS principles,[288] as well as in the deliberations in the Special Political Committee (SPC) of the UNGA. China, for example, took the position that state sovereignty was of utmost importance:

> International direct television broadcasting by satellites should be carried out on the basis of such generally accepted principles of international law as respect for the sovereignty of States, which was an indispensable condition for international co-operation.[289]

This position was shared by Brazil,[290] Hungary,[291] Bulgaria[292] and the USSR, among others.[293] In contrast, the Netherlands, exemplifying the opposite position, reiterated that the freedom of information should be the leading principle.[294]

---

[287] UNCOPUOS (Legal Subcommittee), 'Summary Record of the 290th Meeting' (25 March 1978) UN Doc A/AC.105/C.2/SR.290, 7.

[288] UNGA Special Political Committee (35th Session) 'Summary Record of the 14th Meeting' (27 October 1980) UN Doc A/SPC/35/SR.14, 6.

[289] UNGA Special Political Committee (36th Session) 'Summary Record of the 17th Meeting' (29 October 1981) UN Doc A/SPC/36/SR.17, 7–8.

[290] UNGA Special Political Committee (35th Session) 'Summary Record of the 15th Meeting' (31 October 1980) UN Doc A/SPC/35/SR.15, 7.

[291] UNGA Special Political Committee (35th Session) 'Summary Record of the 15th Meeting' (31 October 1980) UN Doc A/SPC/35/SR.15, 9.

[292] UNGA Special Political Committee (35th Session) 'Summary Record of the 16th Meeting' (3 November 1980) UN Doc A/SPC/35/SR.16, 6.

[293] UNGA Special Political Committee (35th Session) 'Summary Record of the 16th Meeting' (3 November 1980) UN Doc A/SPC/35/SR.16, 16.

[294] UNGA Special Political Committee (35th Session) 'Summary Record of the 17th Meeting' (7 November 1980) UN Doc A/SPC/35/SR.17, 4.

Although this discussion was partially related to the use of outer space for military purposes, it does not inform us on how the mandate of COPUOS was perceived because the discussion did not focus on this aspect. Other statements were delivered, however, that did reflect on the mandate of COPUOS. Chile conveyed its concern that the use of outer space for military purposes was increasing and supported the idea of supplementing the existing rules on the peaceful use of outer space, seeing a 'decisive role' for COPUOS.[295] A similar view was expressed by Austria, which considered one of the main responsibilities of COPUOS to be 'to further the elaboration of fundamental legal principles and norms governing outer space activities'.[296] In connection with the statement that 'Each of those treaties was designed to preserve space as a predominantly peaceful environment',[297] Austria thus acknowledged that COPUOS could discuss legal principles and norms to restrict the military use of outer space. This interpretation was reinforced when Austria discussed recent developments in the USSR and the U.S. that indicated a new phase in space militarisation. Sketching the two possible outcomes, an arms race and conflict in outer space, or peaceful cooperation in outer space, Austria stated that COPUOS 'would have to pay increasing attention to preserving outer space as a peaceful environment'.[298] Sweden supported an Italian proposal that the CD should develop a protocol to the OST on the prevention of an arms race in outer space.[299] Although this seems to indicate that Sweden considered the CD as the appropriate forum to discuss military uses of outer space, Sweden also stated that COPUOS should pay proper attention to the question.[300] Likewise, Brazil stated that negotiations should be initiated within COPUOS to supplement the OST with a protocol to 'preserve outer space as an area free from military activities'.[301]

Other states, such as Romania, the Philippines and the USSR discussed the militarisation of outer space and efforts to limit the use of outer space for military purposes without clarifying whether such efforts should be made in COPUOS or the CD.[302] The USSR also submitted a draft treaty on the prohibition of the stationing of weapons of any kind in outer space but submitted it to the UNGA rather than a

---

[295]UNGA Special Political Committee (35th Session) 'Summary Record of the 14th Meeting' (27 October 1980) UN Doc A/SPC/35/SR.14, 8.
[296]UNGA Special Political Committee (35th Session) 'Summary Record of the 16th Meeting' (3 November 1980) UN Doc A/SPC/35/SR.16, 2.
[297]UNGA Special Political Committee (35th Session) 'Summary Record of the 16th Meeting' (3 November 1980) UN Doc A/SPC/35/SR.16, 2.
[298]UNGA Special Political Committee (35th Session) 'Summary Record of the 16th Meeting' (3 November 1980) UN Doc A/SPC/35/SR.16, 4.
[299]UNGA Special Political Committee (35th Session) 'Summary Record of the 16th Meeting' (3 November 1980) UN Doc A/SPC/35/SR.16, 11.
[300]UNGA Special Political Committee (35th Session) 'Summary Record of the 16th Meeting' (3 November 1980) UN Doc A/SPC/35/SR.16, 11.
[301]UNGA Special Political Committee (36th Session) 'Summary Record of the 16th Meeting' (9 November 1981) UN Doc A/SPC/36/SR.16, 7.
[302]UNGA Special Political Committee (36th Session) 'Summary Record of the 17th Meeting' (6 November 1981) UN Doc A/SPC/36/SR.17, 2 & 6 I UNGA Special Political Committee (36th Session) 'Summary Record of the 16th Meeting' (9 November 1981) UN Doc A/SPC/36/SR.16, 11.

specific committee.[303] This proposal was included in the UNGA agenda and was allocated to the First Committee.[304] Therefore, the discussion on the draft was perceived to fall within the scope of the disarmament framework. Apart from the USSR proposal being discussed in the First Committee, however, no state explicitly stated that efforts to limit the use of outer space for military purposes should be discussed outside COPUOS. The general consensus thus still seemed to be in favour of discussing certain military uses of outer space and, in particular, limitation of the military use of outer space in COPUOS.

The conflict between state sovereignty and the freedom of information was resolved in 1982 through a duty to notify and consult with the receiving state. The final result of the negotiations, Resolution 37/92, indicates that the commercial use of Earth satellites for direct television broadcasting was one of the reasons for adopting the principles and also referred to the potential of the technology to 'have significant international political, economic, social and cultural implications'.[305] Furthermore, and in accordance with the negotiations, the resolution referred to certain concepts that are typically perceived to fall within the military sphere, namely the principle of non-intervention, Principle A(1), and maintaining international peace and security, Principle A(3).

3.2.2.2 Resolution 41/65: Principles Relating to Remote Sensing of the Earth from Space (RS Principles)

Although the Principles Relating to Remote Sensing of the Earth from Space (RS Principles) were adopted 4 years later, the discussion on legal principles pertaining to the remote sensing of the Earth, and in particular to the remote sensing of natural resources on Earth, began at the same time as the discussion on the DBS principles. In 1969, the UNGA adopted the first resolution pertaining to remote sensing.[306] Substantial discussion of the issue in the Legal Subcommittee started in earnest in 1974,[307] after Resolution 3182 (XXVIII) recommended the Legal Subcommittee to consider the legal aspects of remote sensing.[308] Working Group III, responsible for the work on the principles on remote sensing, rather quickly

---

[303]UNGA 'Request for the inclusion of a supplementary item in the Agenda of the Thirty-Sixth Session: Conclusion of a Treaty on the Prohibition of the Stationing of Weapons of Any Kind in Outer Space' UN GAOR 36th Session UN Doc A/36/192 (20 August 1981).

[304]UNGA Special Political Committee (36th Session) 'Summary Record of the 17th Meeting' (6 November 1981) UN Doc A/SPC/36/SR.17, 12.

[305]UNGA Res 37/92 'Principles Governing the Use by States of Artificial Earth Satellites for International Direct Television Broadcasting' (10 December 1982) UN Doc A/RES/37/92.

[306]UNGA Res 2600 (XXIV) (16 December 1969) UN Doc A/RES/2600 (XXIV).

[307]UNGA 'Report of the Committee on the Peaceful Uses of Outer Space' UN GOAR 29th Session Supp No 20 UN Doc A/9620 (1974), 5.

[308]UNGA Res 3182 (XXVIII) (18 December 1973) UN Doc A/RES/3182 (XXVIII).

established certain common elements.[309] Furthermore, a great number of potential legal issues were discussed in the Working Group.

To a large extent, these topics mirrored the issues discussed with respect to the DBS principles. In particular, the subject of prior consent of the sensed state came up, echoing the discussion on the prior consent of the broadcast receiving state. Once more, a large part of the discussion focused on the conflict between the concepts of state sovereignty and freedom of information. Similar to the deliberations on the DBS principles, this issue was partially connected with the use of outer space for military purposes. Moreover, the positions on this issue followed the same split as with the DBS principles: the USSR, Eastern European states and developing states arguing that state sovereignty should have priority and the Western states arguing that freedom of information should be the leading principle.[310] Most relevant for this research, however, is the discussion on the scope of the principles and prohibition of the use of remote sensing data to the detriment of the interests of the sensed state.[311] These topics are the most relevant for this research because the discussion on the scope of the principles gives an indication of the topics that COPUOS can and cannot deliberate on, and the prohibition of the use of the data to the detriment of the interests of the sensed state can include military use.

The subsequent elaboration of the common principles addressed the scope of the principles by tentatively adding the phrase 'of the natural resources of the earth and its environment', thus indicating that the scope of the principles was limited to such remote sensing and excluded the military use of remote sensing (among others).[312] The limitation of the scope of the principles also had ramifications for the prohibition of the use of remote sensing data to the detriment of the interests of the sensed state. The scope already excluded the applicability of the principles to military remote sensing and therefore limited the applicability of the prohibition. The inclusion of this narrow scope in the principles, however, also reduced the relevance of the principles, and the discussion on these principles, to the development of the mandate of COPUOS because the scope put the principles clearly within the mandate of COPUOS. Neither the RS principles nor other discussions in COPUOS around this time had any bearing on the mandate of COPUOS, at least not with respect to the boundary between the mandate of COPUOS and the mandate of the CD.

---

[309] UNCOPUOS, 'Report of the Legal Sub-committee on the work of its Fifteenth Session (3–28 May 1976)' (28 May 1976) UN Doc A/AC.105/171 Annex III, 2–3.

[310] UNGA Special Political Committee (35th Session) 'Summary Record of the 15th Meeting' (31 October 1980) UN Doc A/SPC/35/SR.15, 7 I UNGA Special Political Committee (35th Session) 'Summary Record of the 15th Meeting' (31 October 1980) UN Doc A/SPC/35/SR.15, 9 I UNGA Special Political Committee (35th Session) 'Summary Record of the 16th Meeting' (3 November 1980) UN Doc A/SPC/35/SR.16, 6 I UNGA Special Political Committee (35th Session) 'Summary Record of the 16th Meeting' (3 November 1980) UN Doc A/SPC/35/SR.16, 16 I UNGA Special Political Committee (35th Session) 'Summary Record of the 17th Meeting' (7 November 1980) UN Doc A/SPC/35/SR.17, 4.

[311] UNCOPUOS, 'Report of the Legal Sub-committee on the work of its Fifteenth Session (3–28 May 1976)' (28 May 1976) UN Doc A/AC.105/171 Annex III, 4.

[312] UNCOPUOS, 'Report of the Legal Sub-committee on the Work of its Sixteenth Session (14 March–8 April 1977)' (11 April 1977) UN Doc A/AC.105/196 Annex III, 4–6.

Looking ahead, the same can be said for the major resolutions adopted after the RS principles. Resolution 47/68 stipulates the Principles Relevant to the Use of Nuclear Power Sources in Outer Space that do not pertain to the military use of outer space or the mandate of COPUOS.[313] Resolution 51/122, the Declaration on International Cooperation in the Exploration and Use of Outer Space for the Benefit and in the Interest of All States, Taking into Particular Account the Needs of Developing Countries, supplements the existing treaties and the concept of international cooperation within those treaties.[314] Likewise, Resolution 59/115 on the application of the concept of the launching state,[315] Resolution 62/101 stipulating recommendations on enhancing registration practice,[316] Resolution 68/74 stipulating recommendations on national space legislation[317] and Resolution 62/217 endorsing the Space Debris Mitigation Guidelines[318] encompass recommendations, guidelines and principles that enhance the existing treaties. They do not, however, consider the military use of outer space or limit the military use of outer space. Therefore, these resolutions do not inform the mandate of COPUOS because it is clear that the subject matter falls fully within the scope of COPUOS and does not concern the border between the mandate of COPUOS and the CD.

Accordingly, it will be necessary to turn to other documents to establish the development of the mandate of COPUOS. The documents that will be examined are the yearly UNGA resolutions on international cooperation in the peaceful uses of outer space, including the mandate specified in those resolutions, and the deliberations in the SPC (and later on the Fourth Committee).

3.2.2.3 COPUOS Mandate After the Addition of the 'Prevention of an Arms Race in Outer Space' to the CD Agenda

Following the adoption of the DBS principles, the UNGA decided that COPUOS, apart from considering scientific and technical issues, should focus on the legal

---

[313]UNGA Res 47/68 'Principles Relevant to the Use of Nuclear Power Sources in Outer Space' (14 December 1992) UN Doc A/RES/47/68.

[314]UNGA Res 51/122 'Declaration on International Cooperation in the Exploration and Use of Outer Space for the Benefit and in the Interest of All States, Taking into Particular Account the Needs of Developing Countries' (13 December 1996) UN Doc A/RES/51/122.

[315]UNGA Res 59/115 'Application of the concept of the "Launching State"' (10 December 2004) UN Doc A/RES/59/115.

[316]UNGA Res 62/101 'Recommendations on Enhancing the Practice of States and International Intergovernmental Organizations in Registering Space Objects' (17 December 2007) UN Doc A/RES/62/101.

[317]UNGA Res 68/76 'Recommendations on National Legislation Relevant to the Peaceful Exploration and Use of Outer Space' (11 December 2013) UN Doc A/RES/68/74.

[318]UNGA Res 62/217 'International Cooperation in the Peaceful Uses of Outer Space' (22 December 2007) UN Doc A/RES/62/217 | The guidelines are stipulated in: UNGA 'Report of the Committee on the Peaceful Uses of Outer Space' UN GOAR 62nd Session Supp No. 20 UN Doc A/62/20 Annex (2007).

aspects of remote sensing, the legal norms applicable to the use of nuclear power sources and matters relating to the definition and/or delimitation of outer space and outer space activities.[319] These issues are not relevant to the question whether COPUOS is mandated to discuss the military use of outer space because they clearly fall within the mandate of COPUOS and are not related to the disarmament of outer space. Therefore, these issues do not affect or challenge the mandate of COPUOS (as it relates to the military use of outer space).

In the SPC, however, the potential militarisation of outer space was discussed. It was recalled that COPUOS itself had already urged states to contribute to the prevention of an arms race in outer space.[320] Chile affirmed this by considering COPUOS the appropriate forum to consider the elaboration of the OST to curtail the use of outer space for military purposes.[321] Likewise, the German Democratic Republic (GDR) considered COPUOS the appropriate forum to discuss and prepare agreements on the prevention of the militarisation of outer space.[322] Perhaps most clear was the USSR, which stated:

> COPUOS should continue its useful work in that field and should consider, as a matter of priority, effective measures to prevent the spread of the arms race to outer space.[323]

The USSR took this position because it perceived the disarmament framework as too political and prone to 'sabotage'. This interpretation of the mandate of COPUOS was contradicted by the U.S., which accused the USSR of trying to block consensus on giving to the CD the mandate to discuss the prevention of an arms race in outer space.[324] Moreover, the U.S. saw it as inconsistent that the USSR submitted a draft treaty on the prohibition of the use of force in outer space to the First Committee (charged with disarmament) but then regarded COPUOS as the appropriate forum to discuss such issues.[325] Finally, the U.S. argued that including the prevention of an arms race in outer space in the agenda of COPUOS would gridlock a forum that had been historically productive.[326] In essence, the U.S. thus saw the disarmament framework as the only appropriate forum to discuss the militarisation of outer

---

[319] UNGA Res 37/89 (10 December 1982) UN Doc A/RES/37/89.

[320] UNGA Special Political Committee (38th Session) 'Summary Record of the 18th Meeting' (14 November 1983) UN Doc A/SPC/38/SR.18, 6.

[321] UNGA Special Political Committee (38th Session) 'Summary Record of the 19th Meeting' (8 November 1983) UN Doc A/SPC/38/SR.19, 3.

[322] UNGA Special Political Committee (38th Session) 'Summary Record of the 19th Meeting' (8 November 1983) UN Doc A/SPC/38/SR.19, 10–11.

[323] UNGA Special Political Committee (38th Session) 'Summary Record of the 21st Meeting' (11 November 1983) UN Doc A/SPC/38/SR.21, 8.

[324] UNGA Special Political Committee (38th Session) 'Summary Record of the 21st Meeting' (11 November 1983) UN Doc A/SPC/38/SR.21, 18.

[325] UNGA Special Political Committee (38th Session) 'Summary Record of the 21st Meeting' (11 November 1983) UN Doc A/SPC/38/SR.21, 18.

[326] UNGA Special Political Committee (38th Session) 'Summary Record of the 21st Meeting' (11 November 1983) UN Doc A/SPC/38/SR.21, 19.

space. This interpretation was even more clearly visible in the following Swedish statement:

> At its thirty-seventh session, the General Assembly had requested that the Committee on Disarmament should take action to prevent an arms race in outer space. The Committee on Disarmament was indeed the proper forum for the negotiation of agreements to that end, and the process should begin without delay.[327]

In contrast to these two views that gave the issue to either COPUOS or the CD, Brazil proposed a joint effort between the two:

> A number of proposals had been made recently by France, Italy and the Soviet Union in connection with the Outer Space Treaty or with regard to the prevention of an arms race in outer space. COPUOS was the competent United Nations body to consider ways of filling in the gaps in international legislation concerning outer space. (...)
>
> That did not imply that the negotiation of specific instruments relating to disarmament in outer space should be done in isolation from the Committee on Disarmament, which was the main multilateral negotiating body for questions of disarmament. The work of the two bodies should complement each other.[328]

The inclusion of the prevention of an arms race in outer space in the CD agenda thus resulted in states changing their interpretation or reverting to their previous interpretation of the mandate of COPUOS. Therefore, compared to the mandate identified during the negotiations of the UN space treaties, diverging interpretations of the mandate of COPUOS reappeared. This question was not resolved through Resolution 38/80, which obfuscated the matter by requesting COPUOS to consider questions relating to the militarisation of outer space while taking into account that the CD had been requested to consider the question of preventing an arms race in outer space and the need to coordinate efforts between COPUOS and the CD.[329]

This resolution can be interpreted in two ways. First, it can be taken as confirming all three of the aforementioned interpretations simultaneously because no consensus on the question exists, so the only way to address it in an UNGA resolution is to acknowledge all diverging interpretations. Alternatively, it can be interpreted as a division of mandate between COPUOS and the CD. The paragraph is clear that the CD has the mandate to discuss the prevention of an arms race in outer space. By requesting that COPUOS considers questions relating to the militarisation of outer space, while clearly acknowledging the mandate of the CD, it is implied that COPUOS thus has the mandate to discuss other questions relating to the militarisation of outer space outside the prevention of an arms race.

---

[327]UNGA Special Political Committee (38th Session) 'Summary Record of the 19th Meeting' (8 November 1983) UN Doc A/SPC/38/SR.19, 5 | Italy made a very similar statement in UNGA Special Political Committee (38th Session) 'Summary Record of the 19th Meeting' (8 November 1983) UN Doc A/SPC/38/SR.19, 14.

[328]UNGA Special Political Committee (38th Session) 'Summary Record of the 19th Meeting' (8 November 1983) UN Doc A/SPC/38/SR.19, 4.

[329]UNGA Res 38/80 (15 December 1983) UN Doc A/RES/38/80, par. 15.

Regardless, Resolution 38/80 did not solve the divergence in opinion on where to deliberate upon the questions relating to the militarisation of outer space, and managed to lead it back to COPUOS, where it 'overshadow[ed] the entire twenty-seventh session'.[330] It has been illustrated that the inclusion of questions on the militarisation of outer space had become part of the discussions in COPUOS during the negotiations of the space treaties. Nevertheless, the renewed effort to have an effective disarmament forum and the inclusion of the prevention of an arms race in outer space in the agenda of that forum made the mandate of COPUOS a point of debate again. The then Chairman of COPUOS, Mr. Jankowitsch, clearly stated:

> Although most members shared concern regarding the prospect of the militarisation of outer space, there was no common ground as to the more specific role of the Committee in that regard. There had been very strong disagreement on the mandate of COPUOS in that crucial area of international relations.[331]

In general, the different approaches to the mandate of COPUOS and the CD can thus be divided into three groups: first, the Western group, which regarded the CD as the appropriate forum to discuss questions relating to the militarisation of outer space[332]; second, the USSR, its aligned states and a number of non-aligned states, which considered COPUOS as the appropriate forum to discuss such questions[333]; and, third, some non-aligned states, which deemed it necessary that such questions be dealt with in a cooperative manner.[334] This division is a generalisation to illustrate the three approaches, as both the division of views and the groups are not absolute.

---

[330]UNGA Special Political Committee (39th Session) 'Summary Record of the 39th Meeting' (28 November 1984) UN Doc A/SPC/39/SR.39, 4.

[331]UNGA Special Political Committee (39th Session) 'Summary Record of the 39th Meeting' (28 November 1984) UN Doc A/SPC/39/SR.39, 4.

[332]Exemplified by Sweden: *'The militarization of outer space must be considered in the general context of disarmament. His country firmly supported General Assembly resolution 38/70, which reiterated that the Conference on Disarmament, as the single multilateral disarmament negotiating forum, had a primary role in the negotiation of an agreement on the prevention of an arms race in outer space.'* in UNGA Special Political Committee (39th Session) 'Summary Record of the 42nd Meeting' (30 November 1984) UN Doc A/SPC/39/SR.42, 10 | U.S. in UNGA Special Political Committee (39th Session) 'Summary Record of the 42nd Meeting' (30 November 1984) UN Doc A/SPC/39/SR.42, 9.

[333]Exemplified by a statement made by Vietnam: *'Viet Nam shared the view that COPUOS was competent to consider questions relating to the militarization of outer space and felt that it should be given a more specific mandate in that regard. The prevention of the militarisation of outer space and the peaceful uses of outer space were closely connected.'* in UNGA Special Political Committee (39th Session) 'Summary Record of the 40th Meeting' (29 November 1984) UN Doc A/SPC/39/SR.40, 4 | Czechoslovakia in UNGA Special Political Committee (39th Session) 'Summary Record of the 41st Meeting' (29 November 1984) UN Doc A/SPC/39/SR.41, 8 | Poland in UNGA Special Political Committee (39th Session) 'Summary Record of the 42nd Meeting' (30 November 1984) UN Doc A/SPC/39/SR.42, 3 | USSR in UNGA Special Political Committee (39th Session) 'Summary Record of the 42nd Meeting' (30 November 1984) UN Doc A/SPC/39/SR.42, 13.

[334]Exemplified by the earlier statement by Brazil, but also by Austria: *'Although COPUOS was not a suitable forum for arms control negotiations, it could eventually make a practical contribution to the prevention of an arms race in outer space by supporting negotiations carried out in other*

The USSR, for example, suggested that the CD 'could take up the questions of a material nature and COPUOS could examine the question of political and legal obligations'.[335] Although the weight of the questions would then lie with COPUOS, there would be a role for the CD in the practical execution of the political and legal obligations.

Following the contentious Resolution 38/80, the yearly resolutions on international cooperation in the peaceful uses of outer space adopted a more neutral phrasing on the questions of the militarisation of outer space, namely to consider 'ways and means of maintaining outer space for peaceful purposes'.[336] This is a softening compared to the request to discuss questions relating to the militarisation of outer space and matches better with the initial mandate of COPUOS to consider questions on the use of outer space for peaceful purposes. It did not, however, change the division in the interpretation of the mandate.[337]

In essence, the interpretations of the mandate of COPUOS in the early days of COPUOS reappeared. The USSR argued that COPUOS was the appropriate forum to discuss the prevention of an arms race in outer space because outer space had not yet become an arena of such an arms race.[338] It considered international cooperation in the peaceful use of outer space as closely associated with attainment of the non-militarisation of outer space. In contrast, the Western states considered the prevention of an arms race in outer space a disarmament issue but acknowledged that certain other military uses of outer space were distinct from that issue:

> Most satellites launched hitherto had had military purposes but outer space had remained free from armed conflicts. With the development of specific weapons systems intended for use in outer space, that situation might well be changing.[339]

---

forums.' in UNGA Special Political Committee (39th Session) 'Summary Record of the 41st Meeting' (29 November 1984) UN Doc A/SPC/39/SR.41, 4.

[335] UNGA Special Political Committee (39th Session) 'Summary Record of the 42nd Meeting' (30 November 1984) UN Doc A/SPC/39/SR.42, 13.

[336] UNGA Res 39/96 (14 December 1984) UN Doc A/RES/39/96 I UNGA Res 40/162 (16 December 1985) UN Doc A/RES/40/162 I UNGA Res 41/64 (3 December 1986) UN Doc A/RES/41/64 I UNGA Res 42/68 (2 December 1987) UN Doc A/RES/42/68 I UNGA Res 43/56 (6 December 1988) UN Doc A/RES/43/56 I UNGA Res 44/46 (8 December 1989) UN Doc A/RES/44/46.

[337] UNGA Special Political Committee (40th Session) 'Summary Record of the 38th Meeting' (22 November 1985) UN Doc A/SPC/40/SR.38, 3.

[338] UNGA Special Political Committee (40th Session) 'Summary Record of the 40th Meeting' (25 November 1985) UN Doc A/SPC/40/SR.40, 4.

[339] UNGA Special Political Committee (40th Session) 'Summary Record of the 37th Meeting' (22 November 1985) UN Doc A/SPC/40/SR.37, 5 I Also supported by Australia: *'It was Australia's belief that the military use of satellites did not necessarily run counter to the goal of preserving outer space for peaceful purposes'* in UNGA Special Political Committee (40th Session) 'Summary Record of the 40th Meeting' (25 November 1985) UN Doc A/SPC/40/SR.40, 2.

This is thus a return to the interpretation that there are non-arms military uses (which are peaceful/non-aggressive) and arms military uses (which are non-peaceful/aggressive).

Through the discussions in COPUOS and the SPC, however, progress was made to better define the mandate of COPUOS. By 1989, COPUOS had 'concluded that through its work in scientific, technical, and legal fields it had an important role to play in that area [meaning ways and means of maintaining outer space for peaceful purposes]'.[340] Furthermore, COPUOS concluded that it should focus on the peaceful applications of the achievements of space technology.[341] For all intents and purposes, COPUOS thus chose the approach favoured by the Western states. This means that it would focus on the peaceful applications of space technology (which in the non-aggressive interpretation can still be military as long as it is non-arms) and would support maintaining outer space for peaceful purposes through its work but would not focus on disarmament issues. Nevertheless, certain delegations, such as Kenya,[342] still saw COPUOS as the appropriate forum to discuss the disarmament of outer space. The adoption of Resolution 44/112, however, strongly indicates that a choice in the mandate had been made because in paragraph 5, it states:

> Reiterates that the Conference on Disarmament, as the single multilateral disarmament negotiating forum, has the primary role in the negotiation of a multilateral agreement or agreements, as appropriate, on the prevention of an arms race in outer space in all its aspects.[343]

This does not solve the issue of the mandate pertaining to the non-arms military uses of outer space but does reveal that at least the arms military uses of outer space fall outside the mandate of COPUOS, even the limitation of such uses.

Following Resolution 44/112, an obvious change in general international relations occurred through the end of the Cold War and the subsequent thawing of relations between the West and the East. While the yearly resolutions in the 1990s continued to request COPUOS to consider 'ways and means to maintaining outer space for peaceful purposes',[344] the discussion on the mandate of COPUOS

---

[340] UNGA Special Political Committee (44th Session) 'Summary Record of the 19th Meeting' (27 November 1989) UN Doc A/SPC/44/SR.19, 9.

[341] UNGA Special Political Committee (44th Session) 'Summary Record of the 19th Meeting' (27 November 1989) UN Doc A/SPC/44/SR.19, 9.

[342] UNGA Special Political Committee (44th Session) 'Summary Record of the 21st Meeting' (1 December 1989) UN Doc A/SPC/44/SR.21, 2.

[343] UNGA Res 44/112 (15 December 1989) UN Doc A/RES/44/112.

[344] UNGA Res 45/72 (11 December 1990) UN Doc A/RES/45/72 | UNGA Res 46/65 (9 December 1991) UN Doc A/RES/46/65 | UNGA Res 47/68 (14 December 1992) UN Doc A/RES/47/68 | UNGA Res 48/39 (10 December 1993) UN Doc A/RES/48/39 | UNGA Res 49/34 (9 December 1994) UN Doc A/RES/49/34 | UNGA Res 50/27 (6 December 1995) UN Doc A/RES/50/27 | UNGA Res 51/123 (13 December 1996) UN Doc A/RES/51/123 | UNGA Res 52/56 (10 December 1997) UN Doc A/RES/52/56 | UNGA Res 53/45 (3 December 1998) UN Doc A/RES/53/45 | UNGA Res 54/67 (6 December 1999) UN Doc A/RES/54/67.

subsided. It was acknowledged that such a dispute still existed, but certain states expressed their perspective that Resolution 44/112 had solved the dispute.[345]

However, a further shift occurred in the late nineties. On multiple occasions, the U.S. stated that COPUOS was mandated to exclusively deliberate on peaceful uses of outer space, which excluded non-arms military uses of outer space:

> While other United Nations organs, including the First Committee, held mandates to consider the military uses of outer space, COPUOS offered a forum focused exclusively on promoting the cooperative achievement of benefits from space exploration.[346]

In light of the conclusions drawn in the preceding, the U.S. thus tried to narrow down the mandate of COPUOS. A general understanding had appeared during the negotiations of the space treaties that COPUOS was mandated to discuss the non-arms military uses of outer space, an understanding that had been supported by the U.S. The factual situation, however, seemed quite different. A Chinese statement recounted that COPUOS had discussed, as per its mandate, ways and means of maintaining outer space for peaceful purposes. During those discussions 'the question of the prevention of militarization and of the arms race in outer space had received great attention'.[347] Accordingly, China concluded that COPUOS should continue to devote time to these questions.[348]

The yearly UNGA resolutions on the international cooperation on the peaceful uses of outer space in the 2000s kept the phrasing 'ways and means of maintaining outer space for peaceful purposes'.[349] In general, discussion on the subject was infrequent. Pakistan, for example, stated that the issue of the prevention of the weaponisation of outer space lay within the scope of COPUOS[350] but later argued that the CD and COPUOS should cooperate on the issue.[351] Malaysia likewise

---

[345] UNGA Fourth Committee (49th Session) 'Summary Record of the 20th Meeting' (2 December 1994) UN Doc A/C.4/49/SR.20, 2.

[346] UNGA Fourth Committee (54th Session) 'Summary Records of the 15th Meeting' (17 December 1999) UN Doc A/C.4/54/SR.15, 4 | A similar statement was made the year prior in UNGA Fourth Committee (53rd Session) 'Summary Records of the 11th Meeting' (12 November 1998) UN Doc A/C.4/53/SR.11, 3.

[347] UNGA Fourth Committee (55th Session) 'Summary Records of the 12th Meeting' (21 March 2001) UN Doc A/C.4/55/SR.12, 5.

[348] UNGA Fourth Committee (55th Session) 'Summary Records of the 12th Meeting' (21 March 2001) UN Doc A/C.4/55/SR.12, 6.

[349] UNGA Res 55/122 (8 December 2000) UN Doc A/RES/55/122 | UNGA Res 56/51 (10 December 2001) UN Doc A/RES/56/51 | UNGA Res 57/116 (11 December 2002) UN Doc A/RES/57/116 | UNGA Res 58/89 (9 December 2003) UN Doc A/RES/58/89 | UNGA Res 59/116 (10 December 2004) UN Doc A/RES/59/116 | UNGA Res 60/99 (8 December 2005) UN Doc A/RES/60/99 | UNGA Res 61/111 (14 December 2006) UN Doc A/RES/61/111 | UNGA Res 62/217 (22 December 2007) UN Doc A/RES/62/217 | UNGA Res 63/90 (5 December 2008) UN Doc A/RES/63/90 | UNGA Res 64/86 (10 December 2009) UN Doc A/RES/64/86.

[350] UN Fourth Committee (59th Session) 'Summary Record of the 8th Meeting' (11 November 2004) UN Doc A/C.4/59/SR.8, 9.

[351] UN Fourth Committee (64th Session) 'Summary Record of the 13th Meeting' (24 December 2009) UN Doc A/C.4/64/SR.13, 7.

expressed that such coordination efforts should take place.[352] China acknowledged the problem of the militarisation of outer space but did not expand on the mandate issue.[353] Finally, Russia drew attention to the Russian/Chinese draft treaty on the prevention of the placement of weapons in outer space (PPWT) and alluded that such issues should also be discussed in COPUOS but did not make an explicit statement to that effect.[354]

The enduring discussion,[355] with priority, on ways and means of maintaining outer space for peaceful purposes most recently underwent a change in 2014. The discussion under the item shows the continued presence of previous positions. First, COPUOS should not consider disarmament in, or the weaponisation of, outer space.[356] The second perspective held that COPUOS should deal with certain military uses of outer space.[357] The third position held that COPUOS should deal with disarmament in outer space.[358] Finally, the fourth viewpoint regarded it as necessary that COPUOS cooperates and coordinates with 'other bodies and mechanisms of the United Nations system, such as the First Committee of the General Assembly and the Conference on Disarmament'.[359] These positions are illustrated in Table 3.3.[360]

These perspectives were also apparent in the discussion on the future work of COPUOS.[361] Previously, discussion of the agenda item entitled 'Ways and means of maintaining outer space for peaceful purposes' had led to statements about the inefficiency of the discussions under that item because the Committee's work mostly amounted to 'reaffirmations of allegiance to peace in outer space' and emphasis on

---

[352] UN Fourth Committee (59th Session) 'Summary Record of the 11th Meeting' (23 December 2004) UN Doc A/C.4/59/SR.11, 3.

[353] UN Fourth Committee (59th Session) 'Summary Record of the 7th Meeting' (2 December 2004) UN Doc A/C.4/59/SR.7, 5.

[354] UN Fourth Committee (64th Session) 'Summary Record of the 14th Meeting' (30 November 2009) UN Doc A/C.4/64/SR.14.

[355] UNGA Res 66/71 (9 December 2011) UN Doc A/RES/66/71 I UNGA Res 67/113 (18 December 2012) UN Doc A/RES/67/113 I UNGA Res 68/75 (11 December 2013) UN Doc A/RES/68/75.

[356] UNGA 'Report of the Committee on the Peaceful Uses of Outer Space' UN GOAR 69th Session Supp No 20 UN Doc A/69/20 (1 July 2014), par. 45 and 49.

[357] UNGA 'Report of the Committee on the Peaceful Uses of Outer Space' UN GOAR 69th Session Supp No 20 UN Doc A/69/20 (1 July 2014), par. 43 I Paragraph 43 states that statements have been made to the effect that the right of self-defence under the UN Charter as applied to outer space should be examined, which is clearly a military use of outer space.

[358] UNGA 'Report of the Committee on the Peaceful Uses of Outer Space' UN GOAR 69th Session Supp No 20 UN Doc A/69/20 (1 July 2014), par. 46 and 47.

[359] UNGA 'Report of the Committee on the Peaceful Uses of Outer Space' UN GOAR 69th Session Supp No 20 UN Doc A/69/20 (1 July 2014), par. 48.

[360] The COPUOS report does not name the specific state(s) that made a statement. Therefore, the positions cannot be allocated to specific states.

[361] UNGA 'Report of the Committee on the Peaceful Uses of Outer Space' UN GOAR 69th Session Supp No 20 UN Doc A/69/20 (1 July 2014), 49–51.

3 The Development of the Mandates of the Committee on the Peaceful Uses...

Table 3.3 Interpretation of the mandate of COPUOS in 2014

	First: 'Peaceful' is non-military	Second: 'Peaceful' is non-aggressive	Third: 'Peaceful' includes limiting arms and disarmament	Fourth: COPUOS and CD should co-operate on military use of outer space
Armed military uses	Disarmament framework	Disarmament framework	COPUOS	Cooperation between COPUOS and CD
Non-arms military uses	Disarmament framework	COPUOS	COPUOS	Cooperation between COPUOS and CD
Peaceful uses	COPUOS	COPUOS	COPUOS	COPUOS

the principle of non-militarisation of outer space.[362] As a result, the deliberations on the future work of COPUOS led to the broadening of the agenda item by including as follows:

> [B]roader perspective of space security and associated matters that would be instrumental in ensuring the safe and responsible conduct of space activities, and of identifying effective tools that could potentially provide the Committee with new guidance, in a pragmatic manner and *without prejudice to the mandate of other intergovernmental forums (emphasis added)*.[363]

The inclusion in the agenda item of this broader perspective and the earlier statements illustrate that, despite the divergent perspectives on the mandate of COPUOS, military uses of outer space are, at least in some measure, still considered and deliberated upon in COPUOS. This amended agenda item has been included in the yearly UNGA resolutions since 2014.[364]

Although the agenda item also concerns itself with issues that do not necessarily pertain to the use of outer space for military purposes, such as the long-term sustainability of outer space (LTS), transparency and confidence-building measures, adherence to existing space law and the creation of new standards,[365] a substantial part of the deliberations under the agenda item are about the use of outer space for military purposes and the mandate of COPUOS. The different perspectives on the mandate of COPUOS are still clearly visible in the statements made.[366] Therefore,

---

[362]UNGA 'Report of the Committee on the Peaceful Uses of Outer Space' UN GOAR 69th Session Supp No 20 UN Doc A/69/20 (1 July 2014), 6–7.

[363]UNGA 'Report of the Committee on the Peaceful Uses of Outer Space' UN GOAR 69th Session Supp No 20 UN Doc A/69/20 (1 July 2014), par. 372.

[364]UNGA Res 69/85 (5 December 2014) UN Doc A/RES/69/85 | UNGA Res 70/82 (9 December 2015) UN Doc A/RES/70/82 | UNGA Res 71/90 (6 December 2016) UN Doc A/RES/71/90 | UNGA Res 72/77 (7 December 2017) UN Doc A/RES/72/77.

[365]UNGA 'Report of the Committee on the Peaceful Uses of Outer Space' UN GOAR 72nd Session Supp No 20 UN Doc A/72/20 (27 June 2017), par. 42, 43, 44 & 45.

[366]UNGA 'Report of the Committee on the Peaceful Uses of Outer Space' UN GOAR 73rd Session Supp No 20 UN Doc A/73/20 (5 July 2018), par. 94–97.

the most recent session of COPUOS illustrates that these issues are still heavily discussed and in essence have come full circle. The diverging interpretations that existed at the establishment of COPUOS and seemed to have disappeared during the negotiations of the space treaties have reappeared along the same lines they existed, in the first place.

3.3 The Development of the Mandate of the Conference on Disarmament (CD)

The CD started deliberating on the prevention of an arms race in outer space after it had been given the mandate by the UNGA. In the first resolution dealing with the issue, Resolution 37/83,[367] Article IV OST was reiterated, and it was reaffirmed that 'outer space shall be used exclusively for peaceful purposes and that it shall not become an area for an arms race'. Likewise, Resolution 37/99-D reiterated Article IV OST and emphasised that further effective measures should be adopted.[368]

It is patently clear that the CD has the mandate to discuss the prevention of an arms race and other issues pertaining to disarmament in outer space. It is equally clear that COPUOS has the mandate to discuss issues pertaining to the use of outer space for non-military purposes. The core question concerns the issues that fall between disarmament and the non-military uses of outer space—the non-arms military use of outer space. What has been illustrated in the discussion on the mandate of COPUOS is that COPUOS has discussed the non-arms military use of outer space and, on occasion, even extended its mandate to discuss disarmament in outer space (for example, Article IV OST).

Considering the more specific and more well-described mandate of the CD, it is unlikely that the CD would formally discuss the non-military use of outer space. However, the question then remains whether, in practice, the CD does actually discuss the non-arms military use of outer space. Moreover, might not legal obligations that are adopted to attain disarmament in outer space have a 'spill over' effect on both the non-arms military use of outer space and the non-military use of outer space?[369] This paragraph will examine the deliberations of the CD, and the disarmament framework on the prevention of an arms race in outer space, to answer these two questions.

---

[367]UNGA Res 37/83 'Prevention of an Arms Race in Outer Space' (9 December 1982) UN Doc A/RES/37/83.

[368]UNGA Res 37/99-D 'Prevention of an Arms Race in Outer Space and Prohibition of Anti-Satellite Systems' (13 December 1982) UN Doc A/RES/37/99-D.

[369]Exemplified by the drafts for the PTBT in the Eighteen-Nation Committee which initially contained provisions that would affect the use of rockets, not just those used for the delivery of nuclear weapons or other weapons of mass destruction.

3.3.1 The Prevention of an Arms Race in Outer Space After Resolution S-10/2

That the CD quite strictly kept to its mandate is no surprise. The discussion of the prevention of an arms race in outer space initially considered the stationing of weapons in outer space and the use of outer space for hostile purposes.[370] Although the term 'hostile' is not defined in this context, it leans towards the 'aggressive' or armed military use of outer space. However, there were indications that the mandate of the CD could be more broadly interpreted, such as the use of the phrase 'preventing an arms race in outer space *in all its aspects (emphasis added)*' and the fact that certain states wanted to establish a working group that would determine the relevant issues that the CD would need to consider.[371] Although the former was iterated in Resolution 38/70, the proposal to establish a working group that would consider the issues relevant to the CD was not adopted.[372] In addition, Mongolia submitted a draft treaty on the prohibition of the use of force in outer space and from space against the Earth,[373] which obviously falls within the scope of disarmament and the arms military use of outer space. However, this draft also emphasised the militarisation of outer space, an aspect that can be interpreted more broadly than just the armed military use of outer space.

The emphasis on the militarisation of outer space was reiterated the following year in the statement of the group of socialist states: 'that to prevent outer space from being militarized was a problem of importance of the whole of mankind'.[374] However, the issues that would or would not fall within the concept of militarisation were not clarified. Nevertheless, that the deliberations in the CD did not shy away from issues outside pure disarmament became clear through the propositions made by states on the points that should be discussed in the CD. First, there was the suggestion of strengthening the REG, appealing to states to give more detailed information to ensure easier and better verification.[375] Second, it was proposed to deliberate on a minimum separation distance for satellites in orbit or in transit to

---

[370] UNGA 'Report of the Committee on Disarmament' UN GOAR 38th Session Supp No. 27 UN Doc A/38/27 (6 October 1983), 168–169.

[371] UNGA 'Report of the Committee on Disarmament' UN GOAR 38th Session Supp No. 27 UN Doc A/38/27 (6 October 1983), 167–168.

[372] UNGA Res 38/70 'Prevention of an Arms Race in Outer Space' (15 December 1983) UN Doc A/RES/38/70 I UNGA 'Prevention of an Arms Race in Outer Space: Report of the First Committee' UN GOAR 38th Session UN Doc A/38/633 (10 December 1983).

[373] UNGA 'Conclusion of a Treaty on the Prohibition of the Use of Force in Outer Space and From Space Against the Earth: Report of the First Committee' UN GOAR 38th Session UN Doc A/38/647 (9 December 1983).

[374] UNGA 'Report of the Committee on Disarmament' UN GOAR 39th Session Supp No. 27 UN Doc A/39/27 (2 October 1984), 161.

[375] UNGA 'Report of the Committee on Disarmament' UN GOAR 39th Session Supp No. 27 UN Doc A/39/27 (2 October 1984), 163.

orbit.[376] Further, the idea was advanced that agreement should be sought on establishing cooperative measures to permit the verification of the orbit and the general function of space objects.[377] Finally, it was suggested that the CD should deliberate on a prohibition of damage to, disturbance of or harmful interference in the normal functioning of permitted space objects.[378]

None of these four proposals are pure disarmament issues. Instead, they are focused on supporting disarmament efforts, for example, through enhanced registration practices. Nevertheless, the proposals deal with the non-arms military use of outer space and even the peaceful use of outer space, as is the case with the proposal to seek agreement on a minimum separation distance for satellites. The connection between that suggestion and disarmament in outer space is tenuous. Therefore, just as COPUOS seems to have encroached on the mandate of the CD by discussing non-arms military uses of outer space and disarmament issues, so did the CD encroach on the mandate of COPUOS by deliberating upon peaceful uses of outer space.

Regardless, none of these proposals gained momentum and much of the debate in the CD was on establishing a committee focused on the prevention of an arms race in outer space. This committee was successfully established in 1985 with the mandate 'to examine, as a first step at this stage, through substantive and general consideration, issues relevant to the prevention of an arms race in outer space'.[379] Despite this quite narrow mandate, the CD discussed certain issues that were not directly related to the disarmament of outer space. Although the use of outer space for military purposes was acknowledged as a fact, such use of outer space was also criticised, in particular the use of military satellites for reconnaissance and surveillance, and the use of such satellites to support military operations.[380] Interestingly, one of the proposals for the work of the committee was to 'clarify ambiguities surrounding the existing legal regimes in outer space in terms of what was permitted, what was prohibited, what grey areas might exist and what gaps required attentions'.[381] Even further, it was explicitly stated that the terms 'peaceful purposes' and 'militarisation' had no agreed upon meaning. Once more, states referred to treaties negotiated in COPUOS that could serve the aims pursued in the CD, such as the implementation of the REG as a confidence-building measure to increase

---

[376] UNGA 'Report of the Committee on Disarmament' UN GOAR 39th Session Supp No. 27 UN Doc A/39/27 (2 October 1984), 162.

[377] UNGA 'Report of the Committee on Disarmament' UN GOAR 39th Session Supp No. 27 UN Doc A/39/27 (2 October 1984), 162.

[378] UNGA 'Report of the Committee on Disarmament' UN GOAR 39th Session Supp No. 27 UN Doc A/39/27 (2 October 1984), 163.

[379] UNGA 'Report of the Committee on Disarmament' UN GOAR 40th Session Supp No. 27 UN Doc A/40/27 (3 October 1985), 114.

[380] UNGA 'Report of the Committee on Disarmament' UN GOAR 40th Session Supp No. 27 UN Doc A/40/27 (3 October 1985), 116.

[381] UNGA 'Report of the Committee on Disarmament' UN GOAR 40th Session Supp No. 27 UN Doc A/40/27 (3 October 1985), 117–118.

3 The Development of the Mandates of the Committee on the Peaceful Uses... 97

transparency.[382] The discussions within the Committee thus continued where the discussions in the CD Plenary had left off, discussing not just the disarmament of outer space but also non-arms military uses of outer space.

This continued in subsequent years with the CD, not just discussing non-arms military uses of outer space but even issues that fell much more within the scope of the peaceful use of outer space. With respect to the non-arms military use of outer space, it was acknowledged that outer space was already used in such a manner.[383] Certain states, however, considered the non-arms military use of outer space as contravening the OST, calling for the complete demilitarisation of outer space, including the military satellites.[384] Where the CD delved into the peaceful uses of outer space was with respect to the treaties established by COPUOS, in particular the REG, and in calling for the establishment of an international space agency that would promote international cooperation in the peaceful use of outer space.[385] In addition, the CD discussed the status of outer space as a Common Heritage of Mankind.[386]

In the early nineties, the *ad hoc* Committee still had the same mandate to examine and identify issues, discuss existing agreements and discuss existing proposals and future initiatives pertaining to the prevention of an arms race in outer space.[387] Nevertheless, the CD often deliberated on non-arms military uses of outer space. Certain states acknowledged the use of outer space for military purposes and stated that such use of outer space should be disclosed.[388] Other states argued that the sole

---

[382] UNGA 'Report of the Committee on Disarmament' UN GOAR 40th Session Supp No. 27 UN Doc A/40/27 (3 October 1985), 120.

[383] UNGA 'Report of the Committee on Disarmament' UN GOAR 41st Session Supp No. 27 UN Doc A/41/27 (22 September 1986), 100 I UNGA 'Report of the Committee on Disarmament' UN GOAR 42nd Session Supp No. 27 UN Doc A/42/27 (14 September 1987), 161 I UNGA 'Report of the Committee on Disarmament' UN GOAR 44th Sessions Supp No. 27 UN Doc A/44/27 (22 September 1989), 266.

[384] UNGA 'Report of the Committee on Disarmament' UN GOAR 41st Session Supp No. 27 UN Doc A/41/27 (22 September 1986), 100–102 I UNGA 'Report of the Committee on Disarmament' UN GOAR 42nd Session Supp No. 27 UN Doc A/42/27 (14 September 1987), 161–163.

[385] UNGA 'Report of the Committee on Disarmament' UN GOAR 41st Session Supp No. 27 UN Doc A/41/27 (22 September 1986), 106 I UNGA 'Report of the Committee on Disarmament' UN GOAR 42nd Session Supp No. 27 UN Doc A/42/27 (14 September 1987), 167.

[386] UNGA 'Report of the Committee on Disarmament' UN GOAR 43rd Sessions Supp No. 27 UN Doc A/43/27 (3 October 1988), 216 I UNGA 'Report of the Committee on Disarmament' UN GOAR 44th Sessions Supp No. 27 UN Doc A/44/27 (22 September 1989), 264.

[387] UNGA 'Report of the Committee on Disarmament' UN GOAR 45th Sessions Supp No. 27 UN Doc A/45/27 (21 September 1990), 304 I UNGA 'Report of the Committee on Disarmament' UN GOAR 46th Sessions Supp No. 27 UN Doc A/46/27 (30 September 1991), 274 I UNGA 'Report of the Committee on Disarmament' UN GOAR 47th Sessions Supp No. 27 UN Doc A/47/27 (23 September 1992), 66.

[388] UNGA 'Report of the Committee on Disarmament' UN GOAR 45th Sessions Supp No. 27 UN Doc A/45/27 (21 September 1990), 306–307 I UNGA 'Report of the Committee on Disarmament' UN GOAR 46th Sessions Supp No. 27 UN Doc A/46/27 (30 September 1991), 277.

military use of outer space should be the verification of disarmament.[389] In contrast, other states argued that the CD should work towards the total non-militarisation of outer space, including prohibiting military satellites.[390]

Finally, certain states also discussed the mandate of the CD in relation to the mandate of COPUOS. The argument was put forward that the term 'peaceful' should not be equated with non-aggressive but that it should be interpreted to exclude any military use.[391] This would mean that military uses of outer space fell outside the scope of COPUOS. Moreover, the argument was made that the mandates of the *ad hoc* Committee (and thus the CD) and COPUOS were separate and distinct.[392] However, no elaboration followed this argument. Therefore, it did not illustrate where the non-arms military uses of outer space would fall within the 'separate and distinct' scope of the CD or of COPUOS.

In its last report before the deadlock, the CD discussed a number of issues that clearly went beyond mere disarmament and also illustrated the interrelation between the topics discussed in COPUOS and the CD. First, the CD discussed a Code of Conduct for outer space activities and possible cooperative responses to the proliferation of space debris.[393] Although these issues are related to the disarmament of outer space, they are far from pure disarmament issues. In addition, these issues are also dealt with by COPUOS. Moreover, the CD discussed the overall demilitarisation of outer space,[394] the REG[395] and the concept of '"rules of the road" relating to space debris, manoeuvres in outer space, and the establishment of keep-out zones'.[396]

Naturally, most of the deliberations within the *ad hoc* Committee of the CD on the prevention of an arms race in outer space focused on true disarmament issues such as verification and prohibiting the placement of arms in outer space. This conforms to the yearly UNGA resolutions on the prevention of an arms race in outer space that designates the CD as having the primary role in the negotiation of an

---

[389]UNGA 'Report of the Committee on Disarmament' UN GOAR 45th Sessions Supp No. 27 UN Doc A/45/27 (21 September 1990), 306.

[390]UNGA 'Report of the Committee on Disarmament' UN GOAR 46th Sessions Supp No. 27 UN Doc A/46/27 (30 September 1991), 276.

[391]UNGA 'Report of the Committee on Disarmament' UN GOAR 46th Sessions Supp No. 27 UN Doc A/46/27 (30 September 1991), 277.

[392]UNGA 'Report of the Committee on Disarmament' UN GOAR 46th Sessions Supp No. 27 UN Doc A/46/27 (30 September 1991), 277.

[393]UNGA 'Report of the Committee on Disarmament' UN GOAR 49th Sessions Supp No. 27 UN Doc A/49/27 (26 September 1994), 125.

[394]UNGA 'Report of the Committee on Disarmament' UN GOAR 49th Sessions Supp No. 27 UN Doc A/49/27 (26 September 1994), 127.

[395]UNGA 'Report of the Committee on Disarmament' UN GOAR 49th Sessions Supp No. 27 UN Doc A/49/27 (26 September 1994), 128.

[396]UNGA 'Report of the Committee on Disarmament' UN GOAR 49th Sessions Supp No. 27 UN Doc A/49/27 (26 September 1994), 128.

agreement on disarmament in outer space.[397] However, the preceding has illustrated that the *ad hoc* Committee discussed many non-arms military uses of outer space and even ventured into discussing peaceful uses of outer space.

3.3.2 The Prevention of an Arms Race in Outer Space Since the Deadlock of the CD

In the 1995 session, the CD was not able to come to a consensus on re-establishing the *ad hoc* Committee on the prevention of an arms race in outer space.[398] Although the issue was still being discussed in the plenary meeting of the CD, most states just reiterated their earlier statements.[399] The CD was successful in 1996 in negotiating the Comprehensive Nuclear-Test-Ban Treaty (CTBT).[400] Outer space was included in the treaty because of its broad scope ('not to carry out any nuclear weapon test explosion or any other nuclear explosion, and to prohibit and prevent any such nuclear explosion at any place under its jurisdiction or control' in Article I), but otherwise outer space was not specifically mentioned. Nevertheless, following the adoption of the CTBT, the CD remained deadlocked with no substantive discussions on the prevention of an arms race in outer space.[401]

---

[397] See, amongst others: UNGA Res 41/53 'Prevention of an Arms Race in Outer Space' UN Doc A/RES/41/53 (3 December 1986) I UNGA Res 42/33 'Prevention of an Arms Race in Outer Space' UN Doc A/RES/42/33 (30 November 1987).

[398] UNGA 'Report of the Committee on Disarmament' UN GOAR 50th Session Supp No. 27 UN Doc A/50/27 (22 September 1995), 133.

[399] UNGA 'Report of the Committee on Disarmament' UN GOAR 50th Session Supp No. 27 UN Doc A/50/27 (22 September 1995), 133.

[400] Comprehensive Nuclear-Test-Ban Treaty (adopted 10 September 1996 UNGA Res 50/245, opened for signature 24 September 1996).

[401] Wade Boese, 'CD Deadlock Continues as U.S. and China Square Off' (2000) 30 *Arms Control Today* 26 I UNGA 'Report of the Committee on Disarmament' UN GOAR 51st Session Supp No. 27 UN Doc A/51/27 (12 September 1996), 48–49 I UNGA 'Report of the Committee on Disarmament' UN GOAR 52nd Session Supp No. 27 UN Doc A/52/27 (9 September 1997), 12 I UNGA 'Report of the Committee on Disarmament' UN GOAR 53rd Session Supp No. 27 UN Doc A/53/27 (8 September 1998), 17 I UNGA 'Report of the Committee on Disarmament' UN GOAR 54th Session Supp No. 27 UN Doc A/54/27 (30 September 1999), 9 I UNGA 'Report of the Committee on Disarmament' UN GOAR 55th Session Supp No. 27 UN Doc A/55/27 (21 September 2000), 7 I UNGA 'Report of the Committee on Disarmament' UN GOAR 56th Session Supp No. 27 UN Doc A/56/27 (13 September 2001), 6 I UNGA 'Report of the Committee on Disarmament' UN GOAR 57th Session Supp No. 27 UN Doc A/57/27 (12 September 2002), 7–8 I UNGA 'Report of the Committee on Disarmament' UN GOAR 58th Session Supp No. 27 UN Doc A/58/27 (9 September 2003), 6 I UNGA 'Report of the Committee on Disarmament' UN GOAR 59th Session Supp No. 27 UN Doc A/59/27 (7 September 2004), 7 I UNGA 'Report of the Committee on Disarmament' UN GOAR 60th Session Supp No. 27 UN Doc A/60/27 (22 September 2005), 6–7 I UNGA 'Report of the Committee on Disarmament' UN GOAR 61st Session Supp No. 27 UN Doc A/61/27 (15 September 2006), 6 I UNGA 'Report of the Committee on Disarmament' UN GOAR

During the deadlock, multiple proposals were submitted on how to overcome the deadlock. However, none of these documents resulted in the end of the deadlock. In 2008, China and Russia submitted a draft Treaty on Prevention of the Placement of Weapons in Outer Space and of the Threat or Use of Force against Outer Space Objects (PPWT).[402] The idea for such a treaty was not new; draft treaties on the topic had already been submitted by the USSR in the eighties, but the collaboration on the topic between China and Russia was a new development.

This draft is relevant as it could have implications beyond the disarmament of outer space and even beyond the use of outer space for military purposes. Although the draft stipulates that the definitions in Article I are only for the purpose of the PPWT, the definitions contained within that article can have influence on the further development of definitions of outer space activities.[403] First, the definition of 'outer space' as the 'space above the Earth in excess of 100 km above sea level' could lead to the development of a generally accepted definition of outer space. COPUOS has discussed the delimitation of outer space since 1957 and has not come to an agreed upon definition; thus, it is unlikely that the PPWT would create a uniform definition. Nevertheless, the PPWT could affect this discussion. Likewise, none of the UN Space Treaties has defined the term 'space object'. The PPWT's attempt to do so could have similar ramifications as for the definition of outer space. Therefore, the PPWT could have a bearing on all space activities and not just armed activities in outer space.

At the same time, the European Union (EU) drafted its Draft Code of Conduct for Outer Space Activities (EUCoC).[404] Although Principle 1.4 explicitly stated that adherence to the EUCoC was voluntary, it could have had a similar effect as the PPWT, developing definitions and measures outside the disarmament scope and thus affecting the peaceful use of outer space. Examples of this are the measures on minimising accidents in space and collisions between space objects (Principle 4), space debris control and mitigation (Principle 5) and the notification and registration of space activities and space objects (Principles 6 and 7).

62nd Session Supp No. 27 UN Doc A/62/27 (13 September 2007), 10–11 | UNGA 'Report of the Committee on Disarmament' UN GOAR 63rd Session Supp No. 27 UN Doc A/63/27 (9 September 2008), 9–10 | UNGA 'Report of the Committee on Disarmament' UN GOAR 64th Session Supp No. 27 UN Doc A/64/27 (17 September 2009), 13–14.

[402] UNGA 'Report of the Committee on Disarmament' UN GOAR 63rd Session Supp No. 27 UN Doc A/63/27 (9 September 2008), 9.

[403] CD 'Letter dated 12 February 2008 from the Permanent Representative of the Russian Federation and the Permanent Representative of China to the Conference on Disarmament Addressed to the Secretary-General of the Conference Transmitting the Russian and Chinese Texts of the Draft "Treaty on Prevention of the Placement of Weapons in Outer Space and of the Threat or Use of Force Against Outer Space Objects (PPWT)" Introduced by the Russian Federation and China' UN Doc CD/1839 (29 February 2008).

[404] Council of the European Union, 'Draft Code of Conduct for Outer Space Activities' 17175/08—Annex II (17 December 2008).

Although 2009 saw some success in rekindling the work of the CD through an agreed upon programme of work to break the deadlock,[405] the CD was unable to implement the programme of work and thus still remained in a deadlock on the substantive discussion of agenda items in working groups.[406] This deadlock remains until the present with no substantive discussion of the prevention of an arms race in outer space in an established *ad hoc* working group.[407]

However, that does not mean that the prevention of an arms race in outer space goes wholly undiscussed. Instead, the discussion of the prevention of an arms race in outer space occurs during the general debate in the plenary meeting of the CD.[408] In general, the statements made in the Plenary pertaining to the prevention of an arms race in outer space reiterate earlier positions and focus mostly on calling for negotiations on the topic rather than substantive statements on how to achieve the prevention of an arms race in outer space. Of course, some statements have been made that were more substantive, which continues to illustrate that the CD deliberates more broadly than pure disarmament issues. Russia, for example, has stated that the CD should have as its top priority the prevention of the militarisation of outer space.[409] Likewise, India, on behalf of the Group of 21, has made a statement on the prevention of an arms race in outer space to the effect that the current legal regime needs to be consolidated, reinforced and enhanced to 'deter further militarization of outer space or to prevent its weaponization'.[410]

The further discussion of the EUCoC and the PPWT has also included certain non-disarmament issues. Hungary, on behalf of the EU, for example, has emphasised the risks posed by space debris and the need to minimise interference,

---

[405] Cole Harvey, 'CD Breaks Deadlock on Work Plan' (2009) 39 *Arms Control Today* 42 | CD, 'Draft Decision on the Implementation of CD/1864 for the 2009 Session of the Conference on Disarmament' UN Doc CD/1870/Rev.2 (2 August 2009).

[406] UNGA, 'Report of the Conference on Disarmament' UN GOAR 64th Session Supp No. 27 UN Doc A/64/27 (13 October 2009).

[407] UNGA, 'Report of the Conference on Disarmament' UN GOAR 65th Session Supp No. 27 UN Doc A/65/27 (14 September 2010), 10 | UNGA, 'Report of the Conference on Disarmament' UN GOAR 66th Session Supp No. 27 UN Doc A/66/27 (7 October 2011), 11 | UNGA, 'Report of the Conference on Disarmament' UN GOAR 67th Session Supp No. 27 UN Doc A/67/27 (5 October 2012), 10 | UNGA, 'Report of the Conference on Disarmament' UN GOAR 68th Session Supp No. 27 UN Doc A/68/27 (20 September 2013), 10 | UNGA, 'Report of the Conference on Disarmament' UN GOAR 69th Session Supp No. 27 UN Doc A/69/27 (30 September 2014), 18 | UNGA, 'Report of the Conference on Disarmament' UN GOAR 70th Session Supp No. 27 UN Doc A/70/27 (18 September 2015), 20 | UNGA, 'Report of the Conference on Disarmament' UN GOAR 71st Session Supp No. 27 UN Doc A/71/27 (22 September 2016), 13 | UNGA, 'Report of the Conference on Disarmament' UN GOAR 72nd Session Supp No. 27 UN Doc A/72/27 (22 September 2017), 17 | UNGA, 'Report of the Conference on Disarmament' UN GOAR 73rd Session Supp No. 27 UN Doc A/73/27 (14 September 2018), 10.

[408] *Ibid.*

[409] CD, 'Final Record of the One Thousand One Hundred and Sixty-Fifth Plenary Meeting' UN Doc CD/PV.1165 (2 February 2010), 6.

[410] CD, 'Final Record of the One Thousand One Hundred and Eighty-Eights Plenary Meeting' UN Doc CD/PV.1188 (6 July 2010), 7.

collisions, accidents and the creation of space debris.[411] Furthermore, Switzerland has said:

> It is therefore crucial that we specify what should be permitted and what should be prohibited with regard to the military use of space. It is primarily the responsibility of the Conference on Disarmament to determine how such guidelines should be drafted.[412]

This statement clearly illustrates that the CD mandate is interpreted, at least by some states, as including non-arms military uses of outer space. Nevertheless, most deliberations in the CD have pertained to actual disarmament issues or to the way forward in the CD.

This is still the current situation; the CD deliberates on the PPWT and other issues related to the prevention of an arms race in outer space, but a large part of the deliberations deal with the procedural issues of how to discuss the prevention of an arms race in outer space. The most recent development has been the establishment of the Third Subsidiary Body of the CD on the prevention of an arms race in outer space 'tasked with seeking to reach an understanding on areas of commonalities, deepening technical discussions, and considering effective measures, including legal instruments, for negotiations'.[413] It remains to be seen whether the subsidiary bodies will achieve success in this objective.

3.4 Interim Conclusion on the Mandates of COPUOS and the CD and the Collaboration Between COPUOS and the CD

The previous chapter focused on the initial mandates of COPUOS and of the CD at their respective establishment. The previous two sections in this chapter illustrated the practical interpretation of those mandates and the development of those mandates through the actual deliberations in the two forums.

COPUOS was established with the mandate to discuss the peaceful uses of outer space, where the term 'peaceful' was interpreted in two ways. First, peaceful meant non-military. This interpretation was supported by non-aligned states and, at least at the start of COPUOS, by the USSR and the other Eastern bloc states. The alternative interpretation was that peaceful meant non-aggressive. This interpretation was supported mostly by the U.S. and some of the Western bloc states. These two interpretations of the term 'peaceful' led to different interpretations of the mandate

---

[411]CD, 'Final Record of the One Thousand Two Hundred and Third Plenary Meeting' UN Doc CD/PV.1203 (8 February 2011), 3.

[412]CD, 'Final Record of the One Thousand Two Hundred and Third Plenary Meeting' UN Doc CD/PV.1203 (8 February 2011), 9.

[413]CD, 'Decision' UN Doc CD/2119 (19 February 2018).

of COPUOS. In essence, three interpretations can be discerned, as illustrated in Table 3.1, which is illustrated above.

Very broadly, the U.S. stated on numerous occasions in the beginning of COPUOS that both armed military uses and non-arms military uses of outer space should be discussed in the disarmament framework, with COPUOS having the mandate to discuss the peaceful (meaning non-aggressive) uses of outer space. The USSR was not as consistent in its statements as the U.S. but, in general, advocated the discussion of armed military uses of outer space in the disarmament framework, while non-arms military uses and peaceful uses of outer space should be discussed in COPUOS. This is exemplified by the inclusion of non-arms military uses of outer space in their drafts before COPUOS, such as the use of satellites for espionage and intelligence information gathering. The final interpretation, best exemplified by India, mostly followed the USSR interpretation but added that COPUOS had the mandate to discuss ways to keep the use of outer space for exclusively peaceful purposes. This meant that principles and provisions that are aimed at limiting the use of outer space for exclusively peaceful purposes can be discussed and negotiated in COPUOS.

The further discussions in COPUOS indicated a shift towards the Indian interpretation of the mandate. After all, COPUOS negotiated the OST, which in Article IV does exactly what India described as being within the COPUOS mandate, namely limit the manner in which outer space can be used to ensure that the use of outer space is exclusively for peaceful purposes. In addition, neither the U.S. nor the USSR made strong statements to repudiate the interpretation favoured by India and like-minded states. This interpretation was supported by statements made by Austria, Japan, Czechoslovakia, Hungary, the UAR, Canada, Argentina, Poland and Mexico.[414] President Lyndon B. Johnson even stated that the OST was the 'most important arms control development since the limited test ban treaty of 1963',[415] thereby reaffirming that COPUOS discussed disarmament matters. The further examination of the deliberations of the ARRA, LIAB, REG and MOON supports the conclusion that COPUOS deliberated on non-arms military uses of outer space and disarmament matters.

---

[414] UNGA COPUOS Legal Sub-Committee 'Summary Record of the Fifty-Eight Meeting' (20 October 1966) UN Doc A/AC.105/C.2/SR.58, 4 | UNGA COPUOS Legal Sub-Committee 'Summary Record of the Fifty-Eight Meeting' (20 October 1966) UN Doc A/AC.105/C.2/SR.58, 6 | UNGA COPUOS Legal Sub-Committee 'Summary Record of the Fifty-Eight Meeting' (20 October 1966) UN Doc A/AC.105/C.2/SR.58, 8 | UNGA COPUOS Legal Sub-Committee 'Summary Record of the Fifty-Ninth Meeting' (24 October 1966) UN Doc A/AC.105/C.2/SR.59, 2–3 | UNGA COPUOS Legal Sub-Committee 'Summary Record of the Sixty-Second Meeting' (24 October 1966) UN Doc A/AC.105/C.2/SR.62, 4 | UNGA COPUOS Legal Sub-Committee 'Summary Record of the Sixtieth Meeting' (20 October 1966) UN Doc A/AC.105/C.2/SR.60, 2–3 | UNGA COPUOS Legal Sub-Committee 'Summary Record of the Sixty-Second Meeting' (24 October 1966) UN Doc A/AC.105/C.2/SR.62, 7 | UNGA COPUOS Legal Sub-Committee 'Summary Record of the Sixty-Second Meeting' (24 October 1966) UN Doc A/AC.105/C.2/SR.62, 8.

[415] UNGA First Committee (21st Session) 1492nd Meeting, 428.

However, to conclude that COPUOS is mandated to discuss non-arms military uses of outer space and disarmament issues when those issues are related to limiting the use of outer space for exclusively peaceful purposes is too blunt. First, COPUOS discussed these matters within the context of negotiating general treaties for which they had received a specific mandate by the UNGA, rather than discussing these matters separately. Second, developments within the disarmament framework need to be taken into consideration. The analysis of the Ten-Nation Committee, Eighteen-Nation Committee and the CCD showed that the deliberations and negotiations in those forums were turbulent with deadlocks being common and years between substantive negotiations in the forums or their successors. With such inconsistent discussions within the disarmament framework, it is a natural result that at least some basic provisions on the disarmament of outer space needed to be addressed in some manner, in this case in COPUOS.

The overhaul of the disarmament framework following Resolution S-10/2 should then have led to a clearer distinction between the mandates of COPUOS and the CD. Instead, following this Resolution, there was a reversal of the apparent consensus within COPUOS back to the positions in the years following the establishment of COPUOS, with the addition of a fourth perspective. The previous perspectives were that COPUOS should not consider disarmament in, or the weaponisation of, outer space[416]; that COPUOS should deal with certain military uses of outer space[417]; and that COPUOS should deal with disarmament in outer space.[418] Finally, the fourth perspective regarded it as necessary that COPUOS cooperates and coordinates with 'other bodies and mechanisms of the United Nations system, such as the First Committee of the General Assembly and the Conference on Disarmament'.[419] These interpretations have been set out in Table 3.3, which is illustrated above.

Although there are thus different interpretations of the mandate of COPUOS, the practical discussions within COPUOS illustrate that non-arms military uses of outer space are definitely discussed, and that within the context of the priority agenda item 'Ways and means of maintaining outer space for peaceful purposes' certain disarmament matters are discussed.

At the same time, the CD did not limit itself to disarmament matters but often discussed matters pertaining to the non-arms military use of outer space and even the peaceful use of outer space, for example by discussing the issue of space debris, which is not a disarmament issue. Of course, the creation of space debris through

---

[416] UNGA 'Report of the Committee on the Peaceful Uses of Outer Space' UN GOAR 69th Session Supp No. 20 (1 July 2014) UN Doc A/69/20, par. 45 and 49.

[417] UNGA 'Report of the Committee on the Peaceful Uses of Outer Space' UN GOAR 69th Session Supp No. 20 (1 July 2014) UN Doc A/69/20, par. 43 I Paragraph 43 states that statements have been made to the effect that the right of self-defence under the UN Charter as applied to outer space should be examined, which is clearly a military use of outer space.

[418] UNGA 'Report of the Committee on the Peaceful Uses of Outer Space' UN GOAR 69th Session Supp No. 20 (1 July 2014) UN Doc A/69/20, par. 46 and 47.

[419] UNGA 'Report of the Committee on the Peaceful Uses of Outer Space' UN GOAR 69th Session Supp No. 20 (1 July 2014) UN Doc A/69/20, par. 48.

anti-satellite weapons is a matter that falls wholly within the scope of the CD. The general matter of space debris, however, is an overarching issue for the use of outer space and would more obviously fall within the scope of COPUOS. Other issues that have been discussed in the CD but do not obviously fall within the scope of the disarmament of outer space are the general militarisation of outer space, the strengthening of the REG to ensure better registration practice, minimum separation distances between satellites to reduce the risk of collision, verification of the general function of space objects and the so-called rules of the road. Nevertheless, many states have brought these matters to the attention of the *ad hoc* Committee on the prevention of an arms race in outer space of the CD.

As a result, both COPUOS and the CD discuss matters across the range of types of space activities. Although COPUOS generally deliberates on matters towards the peaceful use of outer space and the CD generally deliberates on matters towards the disarmament of outer space, there is no distinct border between the mandates of the two forums.

In addition to this, the analysis in the preceding sections and chapter illustrated that many of the issues discussed in COPUOS and the CD are interrelated and that certain issues are discussed in both forums. Once more, space debris is the clearest example. It is discussed in COPUOS under agenda item 11 'General exchange of information and views on legal mechanisms relating to space debris mitigation and remediation measures, taking into account the work of the Scientific and Technical Subcommittee'.[420] Likewise, the subject of space debris is often discussed in the CD, for example when it was discussed with respect to the EUCoC or anti-satellite weapons.[421]

Thus, the question arises, how much collaboration is there between the forums on these matters and on space matters in general? When so many of the issues discussed in the forums affect the peaceful use of outer space, the non-arms military use of outer space and the disarmament of outer space alike, one would expect a certain measure of communication and collaboration between the two forums. Such collaboration has been supported by many states already, both in COPUOS and in the CD, for years. For example, Canada states:

> In this regard, Canada argues for security guarantees to be considered by the Conference on Disarmament (CD) and practical safety and sustainability measures for space activities to be considered in the Committee on the Peaceful Uses of Outer Space (COPUOS). To ensure that these forums do not work at odds with one another, increased co-ordination of the CD and COPUOS ought to be given favourable consideration by the Member States of both international bodies.[422]

---

[420] UN COPUOS, 'Report of the Legal Subcommittee on its Fifty-Seventh Session, Held in Vienna from 9 to 20 April 2018' UN Doc A/AC.105/1177 (30 April 2018), 3.

[421] Council of the European Union, 'Draft Code of Conduct for Outer Space Activities' 17175/08—Annex II (17 December 2008) | CD, 'Canada: Working Paper: On the Merits of Certain Draft Transparency and Confidence-Building Measures and Treaty Proposals for Space Security' UN Doc CD/1865 (5 June 2009), 2.

[422] CD, 'Canada: Working Paper: On the Merits of Certain Draft Transparency and Confidence-Building Measures and Treaty Proposals for Space Security' UN Doc CD/1865 (5 June 2009), 4.

Such collaboration, however, has been nearly non-existent. Cooperation exists on a broader UN level through the annual Inter-Agency Meeting on Outer Space Activities (UN-Space). UN-Space, which has been meeting since the mid-seventies and is now in its 38th session, was established 'to promote collaboration, synergy, the exchange of information and the coordination of plans and programmes between the United Nations entities in the implementation of activities involving the use of space technology and its applications'.[423] This objective is quite limited because it is restricted to the 'implementation of activities involving the use of space technology and its applications'. Therefore, UN-Space does not aim at having discussions on the way forward on contentious issues such as how to deal with space debris, space traffic management or keeping the use of outer space for exclusively peaceful purposes. In addition, although the Comprehensive Nuclear-Test-Ban Treaty Organization (CTBTO) is a participating organisation in UN-Space, the CD is not.[424] Thus, although UN-Space is an important collaboration between UN organisations, in particular through their special reports on issues such as space weather, climate change and space for agriculture development and food security,[425] it is not a collaboration in which COPUOS and the CD participate or a collaboration in which progress is made towards political or legal consensus on space matters.

A rare collaboration occurred through the joint panel discussion of the First and Fourth Committees on the possible challenges to space security and sustainability. However, this was a joint panel between the First and Fourth Committee rather than COPUOS and the CD and was convened for the 50th anniversary of the OST.[426] Although the joint panel discussions allowed for a dialogue between the First and Fourth Committees on a number of a space matters relevant to both COPUOS and the CD, it is far from the required collaboration between the two forums that is necessary to adequately resolve those space matters that affect both forums.

---

[423] UN COPUOS, 'Report of the Inter-Agency Meeting on Outer Space Activities (UN-Space) on its Thirty-Seventh Session' UN Doc A/AC.105/1143 (3 April 2018), 1.

[424] UNOOSA, 'Participating Organizations' <http://www.unoosa.org/oosa/en/ourwork/un-space/po.html> Accessed 19 December 2018.

[425] UN COPUOS, 'Space Weather: Special Report of the Inter-Agency Meeting on Outer Space Activities on Developments Within the United Nations System Related to Space Weather' UN Doc A/AC.105/1146 (28 April 2017) | UN COPUOS, 'Space and Climate Change: Special Report of the Inter-Agency Meeting on Outer Space Activities on the Use of Space Technology Within the United Nations System to Address Climate Change Issues' UN Doc A/AC.105/991 (31 March 2011) | UN COPUOS, 'Space for Agriculture Development and Food Security: Special Report of the Inter-Agency Meeting on Outer Space Activities on the Use of Space Technology within the United Nations System for Agriculture Development and Food Security' UN Doc A/AC.105/1042 (8 April 2013).

[426] UNGA Res 71/90, 'International Cooperation in the Peaceful Uses of Outer Space' UN Doc A/RES/71/90 (6 December 2016).

Chapter 4
The Future of the UN Space-Related Framework

Annette Froehlich, Vincent Seffinga, and Ruiyan Qiu

Abstract The overlapping mandates of COPUOS and the CD and the demonstrated need to discuss certain space matters in a cooperative manner, combined with the fact that the collaboration between COPUOS and the CD is extremely limited, raises the question of whether the current UN space-related framework is adequate to deal with current and future space matters. This chapter aims to answer that question. Therefore, it will first give a short overview of some of the more pressing issues that affect the peaceful use of outer space and the disarmament of outer space alike, namely space traffic management (STM), space debris and the long-term sustainability of outer space activities (LTS). Without going into exhaustive detail, this chapter will outline the essence of the subject matter and the need to discuss these matters in a more cooperative effort. Thereafter, this chapter evaluates the current UN space-related framework and how well equipped it is to deal with the three space matters. In light of the deliberations in the COPUOS and the CD, the interrelatedness of space matters and the lack of cooperation between the two forums, the conclusion is drawn that the current UN space-related framework is not able to effectively discuss the challenges posed by near- to medium-term space issues and challenges.

The overlapping mandates of the Committee on the Peaceful Uses of Outer Space (COPUOS) and the Conference on Disarmament (CD) and the demonstrated need to discuss certain space matters in a cooperative manner, combined with the fact that the collaboration between COPUOS and the CD is extremely limited, beg the question whether the current UN space-related framework is adequate to deal with current and future space matters. This chapter will aim to answer that question. Accordingly, it will first give a brief overview of some of the more pressing issues that affect the peaceful use of outer space and the disarmament of outer space alike,

---

A. Froehlich (✉) · V. Seffinga
European Space Policy Institute, Vienna, Austria
e-mail: annette.froehlich@espi.or.at

R. Qiu
International Institute of Air and Space Law, Leiden University, Leiden, Netherlands

namely the ways and means of maintaining outer space for peaceful purposes, space traffic management (STM), space debris and the long-term sustainability of outer space activities (LTS). None of these issues will be described exhaustively. Instead, this chapter will outline the essence of the subject matter and the need to discuss these matters in a more cooperative endeavour (Sect. 4.1). Thereafter, this chapter, as the conclusion of the research, will evaluate the current UN space-related framework and how well it is equipped to deal with the described space matters (Sect. 4.2).

4.1 The Major Space-Related Issues Facing the International Community

4.1.1 Ways and Means of Maintaining Outer Space for Peaceful Purposes

The use of outer space exclusively for peaceful purposes is a topic that has been widely discussed since space matters were first put on the UN agenda following the successful launch of Sputnik I. Resolution 1148 (XII), for example, already refers to the use of outer space exclusively for 'peaceful and scientific purposes'.[1] Furthermore, many other resolutions and legal instruments, including Resolution 1962 (XVIII) and the OST, recognise 'the common interest of all mankind in the progress of the exploration and use of outer space for peaceful purposes'.[2] The recognised importance of keeping outer space exclusively for peaceful purposes led to the inclusion of an obligation under Article IV OST to use the Moon and other celestial bodies exclusively for peaceful purposes.

This obligation, however, has never been extended towards the rest of outer space. Nevertheless, the international community still recognises it as an important objective to maintain outer space for peaceful purposes. Therefore, COPUOS has been discussing 'ways and means of maintaining outer space for peaceful purposes', an agenda item that has been discussed extensively with respect to the development of the mandate of COPUOS in the previous chapter. It bears repeating, however, that the first time the agenda item was included in Resolution 38/80, it requested COPUOS to consider questions relating to the militarisation of outer space while taking into account that the CD had been requested to consider the question of preventing an arms race in outer space and the need to coordinate efforts between

---

[1]UNGA Res 1148 (XII) Regulation, Limitation and Balanced Reduction of All Armed Forces and All Armaments: Conclusion of an International Convention (Treaty) on the Reduction of Armaments and the Prohibition of Atomic, Hydrogen and Other Weapons of Mass Destruction (14 November 1957).

[2]UNGA Res 1962 (XVIII) Declaration of Legal Principles Governing the Activities of States in the Exploration and Use of Outer Space (13 December 1963) | Treaty on Principles Governing the Activities of States in the Exploration and Use of Outer Space, including the Moon and Other Celestial Bodies (adopted 19 December 1966, entered into force 10 October 1967) 610 UNTS 205.

COPUOS and the CD.³ The ensuing discussion on the militarisation of outer space overshadowed the deliberations in COPUOS.⁴ Therefore, the subsequent UNGA resolutions requested COPUOS to consider 'ways and means of maintaining outer space for peaceful purposes'. Although the agenda item is phrased differently, in essence it is still a discussion on the militarisation of outer space and limiting the use of outer space for non-peaceful purposes. This has been demonstrated in the examination of the agenda item in the previous chapter but is also evident when considering that outer space can only be maintained for peaceful purposes when outer space is not used for non-peaceful purposes.

The relation between the discussion in COPUOS on ways and means of maintaining outer space for peaceful purposes and the discussion in the CD on the prevention of an arms race in outer space then also becomes apparent. The prevention of an arms race in outer space is an essential component in maintaining outer space for peaceful purposes because outer space cannot be maintained for peaceful purposes unless an arms race in outer space is prevented. Simply put, the discussion and outcome in COPUOS are dependent on the deliberations and outcome in the CD. Therefore, it is also apparent that these issues cannot be discussed in separate forums but will need to be dealt with in a cooperative manner.

4.1.2 Space Traffic Management (STM)

Space traffic management (STM) is a concept that is currently in development and has no agreed-upon definition. In the Legal Subcommittee, the issue has been discussed since 2016 through the addition of an agenda item on a 'General exchange of views on the legal aspects of space traffic management'.⁵ The inclusion of this agenda item was suggested by Germany, originally as a single issue/item.⁶ The suggested purpose of the item was 'to reflect on the concept of STM, on what it entails and on what consequences it would have for the organization and governance of space activities'.⁷ The proposal further stressed that the item on STM should not

---

³UNGA Res 38/80 'International Co-operation in the Peaceful Uses of Outer Space' (15 December 1983) UN Doc A/RES/38/80, par. 15.

⁴UNGA SPC, 'Summary Record of the 39ᵗʰ Meeting' (28 November 1984) UN Doc A/SPC/39/SR.39, 4.

⁵UN COPUOS, 'Report of the Legal Subcommittee on its Fifty-Fourth Session, Held in Vienna from 13 to 24 April 2015', UN Doc A/AC.105/1090 (30 April 2015), 33.

⁶UN COPUOS Legal Subcommittee, 'Proposal for a Single Issue/Item for Discussion at the Fifty-Fifth Session of the Legal Subcommittee in 2016' on: 'Exchange of Views on the Concept of Space Traffic Management' UN Doc A/AC.105/C.2/2015/CRP.13 (14 April 2015).

⁷UN COPUOS Legal Subcommittee, 'Proposal for a Single Issue/Item for Discussion at the Fifty-Fifth Session of the Legal Subcommittee in 2016' on: 'Exchange of Views on the Concept of Space Traffic Management' UN Doc A/AC.105/C.2/2015/CRP.13 (14 April 2015), 1.

lead to the elaboration on any legal text but should only offer the opportunity to the delegations to have an exchange of views on the subject matter.[8]

Although STM has only recently been added to the agenda of COPUOS, it is not a new concept. Concepts that are extremely similar have been discussed previously in the CD, such as the concept of 'rules of the road' and the establishment of keep-out zones.[9] Outside the UN, the concept has also been discussed, for example through the 2006 Study of the International Academy of Astronautics (IAA), which defined STM as 'the set of technical and regulatory provisions for promoting safe access into outer space, operations in outer space and return from outer space to Earth free from physical or radio-frequency interference'.[10] This definition was reiterated in a further 2018 IAA study on STM.[11] Further studies have been conducted on behalf of the National Aeronautics and Space Administration (NASA)[12] and on behalf of the European Space Agency (ESA).[13]

The re-emergence of the concept can be attributed to the worsening congestion of the space environment and the urgent need for the international community to mitigate the risks of collision between space objects. The addition of STM as a single issue/item on the 2016 agenda of COPUOS invited a number of statements on the concept of STM. For example, the view was expressed that 'the development of a space traffic management regime should be approached by looking at the following elements: the principles contained in the five United Nations treaties on outer space; the corresponding General Assembly resolutions; additional instruments for keeping outer space clean; space debris mitigation; real-time collision avoidance; notifications and confidence-building measures; orbit management and the passage through airspace; and traffic rules in a narrow sense'.[14] Furthermore, the view was expressed that a wide range of space activities would have to be taken into account to ensure effective management of space traffic, including the increased number of small satellites and nano-satellites launched, the initiatives on mega-constellations and

---

[8] UN COPUOS Legal Subcommittee, 'Proposal for a Single Issue/Item for Discussion at the Fifty-Fifth Session of the Legal Subcommittee in 2016' on: 'Exchange of Views on the Concept of Space Traffic Management' UN Doc A/AC.105/C.2/2015/CRP.13 (14 April 2015), 2.

[9] UNGA 'Report of the Committee on Disarmament' UN GOAR 49th Sessions Supp No. 27 UN Doc A/49/27 (26 September 1994), 128.

[10] C Contant-Jorgenson, P Lála, K-U Schrogl, *Space Traffic Management* (International Academy of Astronautics 2006), 17.

[11] K-U Schrogl, C Jorgenson, J Robinson and A Soucek, *Space Traffic Management: Towards a Roadmap for Implementation* (International Academy of Astronautics 2018).

[12] O Brown and others, *Orbital Traffic Management Study: Final Report* (Science Application International Corporation 2016).

[13] R Tüllmann and others, *On the Implementation of a European Space Traffic Management System* (DLR GfR 2017).

[14] UN COPUOS Legal Subcommittee, 'Draft Report' UN Doc A/AC.105/C.2/L.298/Add.1 (7 April 2016), 6.

the active removal of space debris.[15] Other statements were made on STM, but the outcome of the 2016 session of the Legal Subcommittee was that the growing importance of STM was acknowledged and that continuous discussion on this item was necessary.[16]

The discussion on STM continued in 2017, with Germany even stating that the negotiation of a new binding UN treaty might be considered to regulate STM.[17] However, it was also called into question whether it made sense to discuss STM in the Legal Subcommittee because the Legal Subcommittee was perhaps not fully able to analyse major indicators shaping the concepts underlying STM.[18] Nevertheless, the discussion within COPUOS demonstrates that STM is a near- to medium-term future discussion point that will need to be addressed. The statements also illustrate that it is not just an issue that pertains to the civil or peaceful use of outer space but that it also affects the military use of outer space and even the disarmament of outer space, first, because issues such as space debris are indiscriminate whether a space object is used for civil or military purposes and second, because traffic rules and orbit management will be less effective if only non-military space objects are regulated.

Thus, because STM is relevant to space security, the CD has also recognised the urgency of having a regulatory framework for STM. The Deputy Secretary-General of the CD delivered a speech in 2012, asserting that, 'common efforts are urgently needed to address the growing risks and challenges in space security', and there is a 'need to intensify efforts to develop and agree upon common legal standards and rules governing pre-launch notifications, space traffic management and manoeuvres in orbit, and communication between satellite operators'.[19] Furthermore, the CD has a history of discussing matters that are akin to STM, such as the earlier mentioned 'rules of the road' initiative and elements of the EUCoC.

Therefore, it is apparent that STM should be dealt with by COPUOS and the CD in a cooperative manner, first, because STM is already being discussed in both COPUOS and the CD. Second, both COPUOS and the CD have acknowledged the importance and urgency of establishing an STM regime. Finally, an outcome on the topic in COPUOS will have ramifications for the deliberation on the topic in the CD and vice versa.

---

[15] UN COPUOS Legal Subcommittee, 'Draft Report' UN Doc A/AC.105/C.2/L.298/Add.1 (7 April 2016), 7.

[16] UNGA, 'Report of the Committee on the Peaceful Uses of Outer Space' UN Doc A/71/20 (28 June 2016), 34.

[17] UN COPUOS Legal Subcommittee, 'Responses to the set of Questions provided by the Chair of the Working Group on the Status and Application of the Five United Nations Treaties on Outer Space' UN Doc A/AC.105/C.2/2017/CRP.6 (23 March 2017), 5.

[18] UNGA, 'Report of the Committee on the Peaceful Uses of Outer Space' UN Doc A/72/20 (27 July 2017), 29.

[19] Kassym-Jomart Tokayev, Space Security 2012: Laying the Groundwork for Progress (Delivered by Jarmo Sareva on 29 March 2012 at the Annual UNIDIR Conference on Space Security) <https://www.unog.ch/80256EDD006AC19C/(httpSpeechesByYear_en)/5503B6BDA5BA20C2C1257C2200486BA4?OpenDocument> (Accessed 20 December 2018).

4.1.3 Space Debris

Another matter that will have to be addressed by both COPUOS and the CD is the issue of space debris. Space debris has been discussed at length in COPUOS. The Scientific and Technical Subcommittee (STSC) began discussing the issue following UNGA Resolution 48/39 as a specific agenda item, after years of discussing the issue under the general discussion in the Subcommittee and the Committee.[20] Many scientific and technical aspects of space debris were discussed in the STSC with the aim of establishing voluntary space debris mitigation measures.[21] Following the adoption of the Inter-Agency Space Debris Coordination Committee (IADC) Space Debris Mitigation Guidelines, the STSC shifted its focus to the review of those measures.[22] The efforts of the STSC culminated in the adoption of revised space debris mitigation guidelines, adopted by UNGA Resolution 62/217.[23] After the adoption of these guidelines, the STSC continued considering the issue of space debris in light of the relevant developments.[24]

In addition, the Legal Subcommittee added the single issue/item 'General exchange of information on national mechanisms relating to space debris mitigation measures' to its agenda in 2009.[25] Therein the Legal Subcommittee considers the legal aspects of space debris in relation to the established technical and scientific measures because 'the consideration of legal aspects of the undesirable effects of space activities, (...) would become warranted sooner or later'.[26] Since then it has been an annual agenda item of the Legal Subcommittee as a 'General exchange of information and views on legal mechanisms relating to space debris mitigation and remediation measures, taking into account the work of the Scientific and Technical Subcommittee'.[27]

---

[20] UNGA Res 48/39 'International Cooperation in the Peaceful Uses of Outer Space' UN Doc A/RES/48/39 (10 December 1993) | UN COPUOS, 'Report of the Scientific and Technical Subcommittee on the Work of Its Thirty-First Session' UN Doc A/AC.105/571 (10 March 1994), 12.

[21] UNOOSA, *Space Debris Mitigation Guidelines of the Committee on the Peaceful Uses of Outer Space* (UN 2010), iii.

[22] UNOOSA, *Space Debris Mitigation Guidelines of the Committee on the Peaceful Uses of Outer Space* (UN 2010), iv.

[23] UNGA Res 62/217, 'International Cooperation in the Peaceful Uses of Outer Space' UN Doc A/RES/62/217 (22 December 2007), 6.

[24] UNGA, 'Report of the Committee of the Peaceful Uses of Outer Space' UN GOAR 73rd Session Supp No. 20 UN Doc A/73/20 (5 July 2018), 21–22.

[25] UNGA, 'Report of the Committee of the Peaceful Uses of Outer Space' UN GOAR 63rd Session Supp No. 20 UN Doc A/63/20 (20 June 2008), 32.

[26] UNGA, 'Report of the Committee of the Peaceful Uses of Outer Space' UN GOAR 64th Session Supp No. 20 UN Doc A/64/20 (1 July 2009), 29–30.

[27] UNGA, 'Report of the Committee of the Peaceful Uses of Outer Space' UN GOAR 73rd Session Supp No. 20 UN Doc A/73/20 (5 July 2018), 33–34.

The issue of space debris has also been considered by the CD. The deliberations pertaining to space debris carried out in the CD can be split into two discussions. First are the deliberations carried out with respect to initiatives such as the EUCoC. In this regard, the discussion in the CD has focused on avoiding collisions between space objects to minimise the creation of space debris and the general mitigation of space debris. Second, the CD deliberates on space debris in the larger framework of anti-satellite weapon (ASAT) systems. Such weapons, which are aimed at destroying space-based assets, can create an enormous amount of space debris. In the last decade or so, all three major space powers have carried out such tests. China conducted such a test in 2007 by destroying their Fengyun-1C weather satellite.[28] The U.S. conducted a similar operation by shooting down a defunct reconnaissance satellite in 2006 to protect the public from the toxic fuel present in the satellite.[29] Finally, Russia has also shown that it has ASAT capabilities.[30] The CD has deliberated upon treaties limiting ASAT but has never come to conclude such a treaty. However, the discussion on limiting or banning the use of ASAT weapons is still relevant. For example, the absence of provisions adequately addressing ASAT systems in the PPWT is one of the reasons for the U.S. to not accommodate further negotiations on the PPWT.[31]

The fact that space debris affects all uses of outer space indiscriminately determines that it is an issue for both COPUOS and the CD. It is possible for the issue of space debris to be isolated into separate elements that can then be considered by just COPUOS or just the CD, for example the creation of space debris through ASAT tests. As an 'arms military' use of outer space, a prohibition on ASAT tests falls within the mandate of the CD and could be negotiated within that forum. However, there are other aspects of space debris that do not clearly fall within the mandate of either forum, for example active space debris removal or the mitigation of space debris created by military space objects in their normal. Therefore, cooperation between the two forums to resolve the issue comprehensively seems essential.

---

[28] Gene Milowicki and Joan Johnson-Freese, 'Strategic Choices: Examining the United States Military Response to the Chinese Anti-Satellite Test' (2008) 6 Astropolitics 1 | Leonard David, 'China's Anti-Satellite Test: Worrisome Debris Cloud Circles Earth' (Space.com, 2 February 2007) <https://www.space.com/3415-china-anti-satellite-test-worrisome-debris-cloud-circles-earth.html> accessed 21 December 2018.

[29] Eric Hagt, 'The U.S. satellite shootdown: China's response' (*Bulleting of the Atomic Scientists*, 5 March 2008) <https://thebulletin.org/2008/03/the-u-s-satellite-shootdown-chinas-response/> accessed 21 December 2018.

[30] Ankit Panda, 'Russia Conducts New Test of 'Nudol' Anti-Satellite System' (*The Diplomat*, 2 April 2018) <https://thediplomat.com/2018/04/russia-conducts-new-test-of-nudol-anti-satellite-system/> accessed 21 December 2018.

[31] Jeff Foust, 'U.S. Dismisses Space Weapons Treaty Proposal as "Fundamentally Flawed"' (*SpaceNews*, 11 September 2014) <https://spacenews.com/41842us-dismisses-space-weapons-treaty-proposal-as-fundamentally-flawed/> accessed 21 December 2018.

4.1.4 Long-Term Sustainability of Outer Space Activities

A final issue, which in a sense encompasses issues such as STM and space debris, is the long-term sustainability of outer space activities (LTS). Simply put, the goal of LTS as an agenda item of COPUOS is to achieve the sustainability of outer space so that outer space can keep being used and explored. With this goal in mind, it becomes obvious that issues such as space debris and STM are inherent within achieving the long-term sustainability of outer space activities.

Similar to the consideration of space debris, LTS was first added to the agenda of the STSC in 2009.[32] In line with the decision to consider LTS in the STSC, a working group on the long-term sustainability of outer space activities to discuss the issue was established.[33] The working group on LTS had four expert subgroups: (a) Sustainable Space Utilization and Sustainable Development on Earth, (b) Space Situational Awareness (SSA), (c) Weather and (d) Regulatory Regime and Guidance for Actors in Space. A first set of LTS guidelines was agreed in 2016,[34] with consensus being found on a preamble and nine additional guidelines in 2018.[35] Nevertheless, the mandate of the Working Group came to an end without COPUOS coming to a consensus on the LTS guidelines.[36] Considering the importance of the long-term sustainability of outer space activities, it is unlikely that this will be the end of the discussion in COPUOS.

The discussion on the long-term sustainability of outer space activities has mostly been confined to COPUOS. Nevertheless, it is an issue that is a concern not just for COPUOS but for the CD as well, first, because military uses of outer space and in particular arms used in outer space heavily contribute to the non-sustainability of outer space activities through the creation of space debris and interference with outer space activities. Second, the degradation of the space environment and the non-sustainability of outer space activities heavily affect the military use of outer space. Similar to the issue of space debris, the degradation of the space environment is indiscriminate and will pose the same problems to the military use of outer space as it will to the non-military use of outer space. The uses of outer space considered in the CD thus both contribute to the non-sustainability of outer space activities and are affected by the consequences of the degradation of the space environment. Therefore, it seems sensible that the long-term sustainability of outer space activities is not

---

[32] UNGA, 'Report of the Committee on the Peaceful Uses of Outer Space' UN GOAR 64th Session Supp No. 20 UN Doc A/64/20 (1 July 2009), 21.

[33] UN COPUOS, 'Report of the Scientific and Technical Subcommittee on its Forty-Seventh Session, Held in Vienna from 8 to 19 February 2010' UN Doc A/AC.105/958 (11 March 2010), 26.

[34] UNGA, 'Report of the Committee on the Peaceful Uses of Outer Space' UN GOAR 71st Session Supp No. 20 UN Doc A/71/20 (28 June 2016), Annex.

[35] UN COPUOS, 'Report of the Scientific and Technical Subcommittee on its Fifty-Fifth Session, Held in Vienna from 29 January to 9 February 2018' UN Doc A/AC.105/1167 (6 April 2018), Annex III.

[36] UNGA, 'Report of the Committee on the Peaceful Uses of Outer Space' UN GOAR 73rd Session Supp No. 20 UN Doc A/73/20 (7 July 2018), 27.

just discussed in COPUOS or in the CD. Instead, it seems necessary to discuss the subject in a more cooperative manner.

4.2 The Future Relevance of the UN Space-Related Framework

In light of the aforementioned, it is apparent that there are issues in outer space that need to be addressed on an international level (*e.g.* space debris, STM and LTS). These issues affect the military and non-military use of outer space alike. Space debris is indiscriminate as regards the general function of the space object it hits, STM will need to take into consideration all space objects to be effective and the long-term sustainability of outer space activities cannot be achieved if only non-military space activities adhere to the guidelines. Furthermore, it is evident that decisions taken in one forum will have consequences for the deliberations in the other forum. For example, an obligation for states to mitigate the creation of space debris negotiated in the CD will have ramifications for non-military space activities and alter the discussion on that subject in COPUOS. COPUOS will need to take the obligation created under the auspices of the CD into consideration in its further deliberations on the subject and might be limited in creating further obligations on space debris mitigation. Likewise, an instrument on STM negotiated in COPUOS will have ramifications for the discussions on a minimum separation distance between satellites or 'rules of the road' undertaken in the CD.

This then leads to the question of whether space matters are currently being discussed effectively internationally and whether changes are necessary to improve the discussion of issues pertaining to the use of outer space. Of course, this also leads to the question of what it means for COPUOS and the CD to effectively deliberate on space matters. Is it sufficient that the forums can annually discuss space matters in a cooperative manner, or is it necessary that the forums come to some manner of conclusion on the issues discussed, whether that is in the form of an international agreement, UNGA resolution, or guidelines? Considering that the objective of both the CD and COPUOS is to achieve tangible success in their respective purposes, effectiveness must be seen as the latter; a conclusion or result should be reached on the matters discussed in the forum. That result does not have to be reached within a certain period of time or in a certain form, but it is evident that a forum that does not achieve any modicum of success in producing a result is not effective.

In the current situation, it is a fair assessment that the deliberations on the prevention of an arms race in outer space in the CD have been far from effective since the deadlock in the mid-nineties. No tangible results have been achieved in developing legal, binding or non-binding, instruments to ensure the prevention of an arms race in outer space. Although China and Russia have submitted the PPWT draft, substantive discussions on that draft have been scarce, if not completely absent. The absence of political will to re-establish the *ad hoc* committees of the

CD, including the committee on the prevention of an arms race in outer space, has resulted in hardly any substantive discussions at all.

In contrast, COPUOS has always been a fairly successful international forum. It has demonstrated its success through the negotiation of the five UN Space Treaties, and, in the subsequent absence of consensus on having legally binding instruments, has continued to successfully negotiate resolutions on a variety of issues. One can argue that the non-legally binding resolutions are a step back in the effectiveness of the forum because they lack a legally binding character and might therefore not be as persuasive and efficient in achieving their purposes. The fact remains, however, that states discuss a variety of space matters annually in COPUOS. Although they often disagree on the approach, development and necessary outcome on the matters, progress is made, compromises are reached and results are achieved.

How does this reflect on the discussion of the aforementioned space matters? In its current form, the UN space-related framework consists of the CD, which is in a deadlock, and COPUOS, which adopts non-legally binding instruments. As long as the CD remains in a deadlock, it is unlikely that any effective progress will be made in that forum. Moreover, the length of the deadlock in the CD and the limited progress that has been made in resolving the deadlock suggest that a breakthrough of the deadlock is an unlikely event. Therefore, it is reasonable to expect that any progress on the issues of space debris, LTS and STM will need to come from COPUOS. Considering the diverging interpretations on the mandate of COPUOS, however, it is unlikely that COPUOS will address the military aspects of space debris, LTS and STM. The current UN space-related framework thus cannot adequately deliberate on these space matters.

Therefore, it is necessary that changes are made to the UN space-related framework. Although some progress could be made if states were to agree to re-establish the committees of the CD, this would be only limited progress, given that results in one forum affect the deliberations in the other forum and that space matters cannot simply be divided between COPUOS and the CD.

Instead, what is necessary for the effective discussion of space matters internationally, with the aim of achieving tangible results, is consistent cooperation between the two forums. This is made apparent by the inherent dual use nature of space technology and the demonstrated interrelated nature of the issues. Cooperation between COPUOS and the CD would allow the international community to discuss all facets of space matters and not be limited to the discussion of one facet when the issue cannot be resolved effectively without discussing them all. However, two important notes need to be placed. First, this cooperation can only effectively exist when the CD resolves its deadlock. Second, such cooperation runs the risk of falling in the same pitfall as the CD, namely that states cannot agree on how to proceed with the discussion of the disarmament aspects of certain space matters, which will then lead to a deadlock. This is a reasonable expectation because cooperation between the CD and COPUOS will bring with it the history of the discussions conducted in those two forums. The difficulties encountered in having effective discussions in the CD will thus be carried over to the cooperative effort. In addition, the military use of outer space has proven to be a difficult, political topic to discuss. When COPUOS was requested by the UNGA to discuss the militarisation of outer space and the need to coordinate efforts between COPUOS and the CD,[37] the topic overshadowed the entire session and prevented COPUOS from having effective substantive discussions.[38] Historic evidence thus indicates that a cooperative effort between COPUOS and the CD could grind to a halt because of the contentious and political nature of the topic of the military use of outer space.

An alternative approach would be to take the example of the deliberations on the OST. Within the context of the OST, COPUOS deliberated on the limitation of the use of outer space to ensure that the use of outer space is for exclusively peaceful purposes. Essentially, Article IV OST is a disarmament measure negotiated in COPUOS. Likewise, the UNGA could bestow a specific mandate on COPUOS to discuss disarmament matters within the context of certain space matters, for example by giving a specific mandate to COPUOS to discuss anti-satellite weapons within the context of space debris mitigation. However, the same problem arises as with having stronger, consistent cooperation between the two forums. The space matters to be discussed will bring with them the complications faced in the CD, which can then taint the discussions in COPUOS and even lead to a deadlock in COPUOS.

The solutions are thus not without their problems. Nevertheless, one thing remains certain: space technology will continue to develop, and this development will bring with it new legal problems that will need to be addressed in a more timely and effective manner than is currently the case.

---

[37] UNGA Res 38/80, 'International Co-operation in the Peaceful Uses of Outer Space' (15 December 1983) UN Doc A/RES/38/80, par. 15.
[38] UNGA SPC 'Summary Record of the 39th Meeting' (28 November 1984) UN Doc A/SPC/39/SR.39, 4.

CPSIA information can be obtained
at www.ICGtesting.com
Printed in the USA
LVHW080828150320
650074LV00003B/208